Contents

The Next Age of Disruption

The Digital Future of Management Series from
MIT Sloan Management Review
Edited by Paul Michelman

How to Go Digital: Practical Wisdom to Help Drive Your Organization's Digital Transformation

What the Digital Future Holds: 20 Groundbreaking Essays on How Technology Is Reshaping the Practice of Management

When Innovation Moves at Digital Speed: Strategies and Tactics to Provoke, Sustain, and Defend Innovation in Today's Unsettled Markets

Who Wins in a Digital World? Strategies to Make Your Organization Fit for the Future

Why Humans Matter More Than Ever

How AI Is Transforming the Organization

A Manager's Guide to the New World of Work: The Most Effective Strategies for Managing People, Teams, and Organizations

The Next Age of Disruption

The Next Age of Disruption

MIT Sloan Management Review

The MIT Press
Cambridge, Massachusetts
London, England

This book was set in Stone Serif and Stone Sans by Jen Jackowitz. Printed and bound in the United States of America.

Library of Congress Cataloging-in-Publication Data

Names: MIT Sloan Management Review, compiler.
Title: The next age of disruption / MIT Sloan Management Review.
Description: Cambridge, Massachusetts : The MIT Press, 2021. | Series: The digital future of management | Includes bibliographical references and index.
Identifiers: LCCN 2020018951 | ISBN 9780262542210 (paperback)
Subjects: LCSH: Technological innovations. | New products. | Leadership. | Organizational change.
Classification: LCC HD45 .N458 2021 | DDC 658.4/063--dc23
LC record available at https://lccn.loc.gov/2020018951

10 9 8 7 6 5 4 3 2 1

Series Foreword

Books in the Digital Future of Management series draw from the print and web pages of *MIT Sloan Management Review* to deliver expert insights and sharply tuned advice on navigating the unprecedented challenges of the digital world. These books are essential reading for executives from the world's leading source of ideas on how technology is transforming the practice of management.

Paul Michelman
Editor in chief
MIT Sloan Management Review

Introduction: Is Your Organization Prepared for What's Next?

Karen Dillon

When Clayton M. Christensen and I first appeared at an event to discuss a book that we had coauthored, I wasn't surprised to see a crowd gathering to talk to him after we spoke. The opportunity to share an observation with or ask a question of Clay, one of the world's truly original thinkers, was special. But I had to laugh at how often that moment turned into a request for a selfie with him—Clay had clearly become a rock star of management thinking. He was patient with every photo request, grateful that a new generation was interested in his ideas, and truly eager to learn how those ideas were being used and advanced.

I was honored to be asked by the editors of *MIT Sloan Management Review* to guest-edit this collection. Having spent the past decade working closely with Clay, I'd had the chance to explore a wide range of relevant topics with him, including how the coming of AI, the rapid speed of innovation, and easy access to capital will affect how companies compete in the years ahead—all topics I discussed with him and we, in turn, explore here. I started working with the team at *MIT SMR* before Clay's death in January 2020, so I was able to sit down with him for what would be his last interview; see "Disruption 2020: An Interview with Clayton M. Christensen," which follows this introduction.

Clay had continually refined his own theories over the years, but he was still wrestling with many questions, as you'll see in our Q&A. He was thrilled, too, to learn how the academics and practitioners featured in this collection were thinking about some of the biggest innovation challenges looming for companies. Some of the authors here are former students or collaborators of his; others are simply people whose work Clay admired and followed. For Clay, the hunt for answers—wherever they may be found—was part of his DNA. He was more interested in getting to the right answer than in being right. And for him, getting to the right answer started with asking the right questions. So that's what we focused on in this book. The rules of competition are changing—again. *Is your organization prepared for what's next?*

To help you answer that question, we have broken the book into three sections, each of which addresses challenges that companies continue to encounter:

1. The human element: What role will people play in shaping or responding to disruption? Decades after Christensen first helped us understand how disruption occurs, why have companies not gotten better at solving the innovator's dilemma? Scott D. Anthony (one of Clay's long-time collaborators) and Michael Putz address this critical question in "How Leaders Delude Themselves about Disruption." The answer, they believe, in part lies in leaders' ignorance to disruption occurring in their own industry.

But even the best disruptive ideas don't always make it to market. Roadblocks emerge within companies and with potential investors long before consumers have a chance to vote. Call it the Innovator's Paradox. Jeff Dyer, Nathan Furr, and Mike Hendron offer hope, exploring why good ideas get stuck and how innovators can break through skittish potential supporters' concerns.

In an age of rapid innovation, are companies forgetting their obligation to protect the public trust? Predicting where your industry will stumble within this new world can make the difference in ensuring your company flourishes with its reputation intact. SAP's chief innovation officer, Max Wessel (a former student of Clay's), and his colleague Nicole Helmer explore one of the most important ethical issues in an age of rapid innovation.

And in a special focus on the workforce of the future, we explore what skills tomorrow's employees will need, how workforce education is itself in the process of being disrupted, and Amazon's groundbreaking employee education plans. In the face of rapid technological changes like automation and artificial intelligence, helping employees keep pace is challenging. And companies are wrestling with how to retain top talent—a critical differentiator in a hypercompetitive environment. Companies can no longer afford to wait for the traditional "system" to supply the workers they hope will help shape their future—the need is too acute and too urgent. In three separate articles, we explore here how the workforce of the future will need to be shaped to compete.

2. The competitive element: Where will competitive advantage be gained (or lost) by understanding how new disrupters are evolving? Columbia Business School's Rita Gunther McGrath, whose work Christensen long admired, explores a new breed of disrupters that are playing by different rules. Companies such as Warby Parker, Dollar Shave Club, Glossier, Away, Casper, and

Bonobos are upending categories as varied as men's grooming, travel, home furnishings, and clothing. Their value proposition to customers almost always features a comparable product at a lower price. More importantly, they always offer a shopping experience that eliminates many of the frictions and irritants of conventional retail. What does the ascendance of a new generation of straight-to-consumer disrupters tell us about how innovation is evolving? McGrath offers her insights here. And in a separate piece, HubSpot CEO Brian Halligan explores how a disruptive product is not sufficient anymore—companies need a disruptive customer experience as well.

Marco Iansiti and Karim R. Lakhani, Christensen's colleagues at Harvard Business School, discuss how reinventing a company around data, analytics, and artificial intelligence changes the game, removing constraints on scale, scope, and learning that have restricted business growth for hundreds of years. Traditional businesses across the economy are being attacked by highly scalable, data-driven companies that harness AI for competitive advantage. Is your company prepared to compete in the age of AI?

MIT's Michael A. Cusumano, Harvard's David B. Yoffie, and University of Surrey's Annabelle Gawer share their thinking on how platforms can power some of the world's most valuable companies. But that alone, they argue, won't necessarily allow them to harness their potential to their disruptive potential for competitive advantage.

3. The futurist element: What do we need to understand about where disruptive threats will emerge in the future? Futurist Amy Webb explores the 11 sources of disruption every company must monitor—and they're not as obvious as you might imagine.

When faced with deep uncertainty, teams often develop a habit of controlling for internal, known variables and forget to track external factors as potential disrupters, Webb explains. Thinking too narrowly can leave your company unprepared for the emergence of powerful challengers. This is a must-read for everyone.

Academics Joshua Gans (Rotman) and Rahul Kapoor (Wharton) and Thomas Klueter (IESE) explore the complexities of charting a course of disruption. Christensen's classic playbook on disruption, they suggest, is not the only strategy companies need. As Gans explains, disruption is a *choice*—but it's not the only one that will lead to success. Separately, Kapoor and Klueter discuss the necessity of working uncertainty into any disruptive value proposition to prepare for what you can't anticipate.

Even in an age of open innovation and global collaboration, there isn't an "add water and stir" approach to outsourcing innovation. Companies must be prepared to partner with both external and internal innovators to gain competitive advantage. Neil C. Thompson, Didier Bonnet, and Yun Ye explain why this blended approach is essential.

As technology and capital rapidly increase the pace of innovation, Christensen's thinking is more relevant today than ever. What do we know about the power of disruption and where it's taking us? That's still an important question to answer. As Christensen himself discusses in his last interview, published here: "Companies certainly know more about disruption than they did in 1995, but I still speak and write to executives who haven't firmly grasped the implications of the theory. The forces that combine to cause disruption are like gravity—they are constant and are always at work. . . . It takes very skilled and very astute leaders to be navigating disruption on a constant basis."

We tackle all these questions in this book, dedicated to the memory of the original disrupter, Clayton M. Christensen. And we urge you to keep asking good questions on your own. The future, Christensen believed, is for those who dare to ask hard questions, even when easy answers are not to be found.

I

The Theory of Disruption

1

Disruption 2020: An Interview with Clayton M. Christensen

Clayton M. Christensen, interviewed by Karen Dillon

In the decades since Clayton M. Christensen first shared his Theory of Disruptive Innovation with the world, his thinking has led to the creation of billions of dollars of revenue, hundreds of companies, and an entirely new paradigm for how industry entrants upend established giants. Karen Dillon, Christensen's longtime collaborator, had a chance to sit down with him before his death in January 2020 to learn how he had refined his thinking, what the future of innovation looked like through that lens, and what questions he was still wrestling with. This is an edited version of their conversation.

***MIT Sloan Management Review*: Over the years, the phrase *disruptive innovation* has come to mean all manner of things to people. But the broad, sweeping implication that "disruptive" is synonymous with "ambitious upstart" is not correct, is it? How would you like to define disruptive innovation for the record?**

Clayton M. Christensen: Disruptive innovation describes a process by which a product or service powered by a technology enabler initially takes root in simple applications at the low end of a market—typically by being less expensive and more accessible—and then relentlessly moves upmarket, eventually displacing established competitors. Disruptive innovations are not breakthrough innovations or "ambitious upstarts" that dramatically alter how business is done but, rather, consist of products and services that are simple, accessible, and affordable. These products and services often appear modest at their outset but over time have the potential to transform an industry. Robert Merton talked about the idea of "obliteration by incorporation," where a concept becomes so popularized that its origins are forgotten. I fear that has happened to the core idea of the theory of disruption, which is important to understand because it is a tool that people can use to predict behavior. That's its value—not just to predict what your competitor will do but also to predict what your own company might do. It can help you avoid choosing the wrong strategy.

You have been a big proponent of the benefits of causal theory. What do you think of the argument that big data obviates the need to seek causality?

Well, it's important to first recognize that the data are not the phenomena. They are a *representation* of the phenomena. Also, we must recognize that God did not create data; any piece of data you or I have ever encountered was created by a human being. Unable to fully capture this wonderfully complex world, we human beings use our bounded rationality to make "decisions" about what aspects of the phenomena to include, and which to exclude, in our data.

These decisions become embedded in the tools we use to create and process data. By definition, these decisions reflect our preexisting ways of thinking about the world. These ways of thinking are sometimes good and reliable—guided by known causal relationships. But oftentimes they are not. No quantity, velocity, or granularity of data can solve this fundamental problem.

I believe that in order for our scientific understanding of the world to progress, we must continually crawl inside companies, communities, and the lives of individuals to create new data in new categories that reveal new insights.

As an example, in my early research on the disk drive industry, I catalogued by hand every disk drive that had been bought or sold over the years after scouring hundreds of "Disk/Trend" reports. And while I was starting to see a pattern of the low-end companies quickly rising to prominence and challenging established leaders, it wasn't until I went out to Silicon Valley and spoke with executives in the space that I fully grasped how incapable incumbent leaders are of responding to disruptive entrants. The data alone would have never generated those insights.

Big data also tends to gloss over or ignore anomalies unless it's crafted carefully to surface these to humans. That is, big data tends to be far more focused on correlation rather than causation and as such ignores examples where something doesn't follow what tends to happen on average. It's only by exploring anomalies that we can develop a deeper understanding of causation. If you think about it, following a big-data approach is what powered our understanding of the sun, moon, stars, and Earth for years, but it was only when Galileo peered through a telescope that we could start to understand more deeply how these celestial bodies moved in relation to one another.

You have commented that the inability to create disruptive growth helps explain Japan's economic malaise. Do you worry that the series of mergers resulting in bigger and bigger com-

panies that seem to primarily focus on stock buybacks is creating the same conditions for the United States?

I absolutely worry about this. In the latest book that you and I wrote together, *The Prosperity Paradox*, we describe three types of innovation, all of which have a different impact on the growth of a firm and—by extension—a nation. *Sustaining innovation*, which most understand, is the process of making good products better. This is important for any economy, but once a market is mature, it generates little net growth in terms of new factories, new jobs, new technology investments, and so forth. There is also *efficiency innovation*, which is when a company tries to do more with less. By their very nature, efficiency innovations don't create new growth, because their purpose is to squeeze more out of what you're putting in. They generate free cash flow for companies, which is important, but if not reinvested properly, that cash doesn't necessarily lead to new growth. A third type of innovation consists of developing simple products for unserved populations who historically couldn't afford or didn't have access to something. These are what we call *market-creating innovations*, meaning they build a new market for new customers. These innovations are the source of growth in any economy, as they pull in resources, investment, operations, employees, and infrastructure in order to serve this larger population of customers.

My sense is that we in the United States, like many other developed countries, are investing far too much energy in efficiency and sustaining innovations, and not enough in market-creating innovations. Buybacks are not inherently wrong, but at an extreme they indicate an inability of a firm (and perhaps an entire economic system!) to identify market-creating opportunities. There are many reasons why this is occurring, but despite some recent incremental improvements to GDP and unemployment, the long-term economic picture doesn't seem too rosy to me as long as this more fundamental problem goes unaddressed.

In 2013, you made an off-the-cuff prediction that 50% of the 4,000 colleges and universities in the United States would go bankrupt in 10 to 15 years. I know that, at the time, you were saying that in a spontaneous conversation, but this observation has been cited many times since as the "doomsday knell" of higher education. Now that you've had more time to think through this prediction, do you want to revise it?

I'll clarify a few things about the prediction. Rather than focus on bankruptcy, which is hard for colleges to declare (for regulatory reasons), what we'll ultimately see is a lot of college closures and mergers. Since 2015, 14 schools have closed and nine have merged in New England alone. A new consulting firm was recently developed to help colleges merge. So this problem is not going away. I think 50% is on the high end of the scale, but not out of the realm of possibility, and 25% to 30% of colleges failing over the next couple of decades is very realistic.

My colleagues have been extremely insightful and have added enormous precision and insight to what I predicted many years ago. Michael Horn, one of my coauthors on *Disrupting Class* and a cofounder of the Clayton Christensen Institute, has recently written a very detailed summary of what in reality was a prediction of 25% that we made together in *The New York Times* in 2013. Although disruption—in the form of faster, more affordable, and more convenient college alternatives powered by online learning—is accelerating and a huge threat to established institutions, ultimately I've always felt that the bigger imminent danger is that their business models simply aren't sustainable.

We'd love to hear your thoughts on the nature of disruption today versus two decades ago. How has the threat to incumbents evolved? How has the opportunity to disrupt established markets transformed? We assume that everything has sped up and that the threats of displacement are greater today – but is that really so?

The mechanics of disruption are the same as ever, but recent technological and business model innovations present unique opportunities and challenges for both incumbents and entrants. For example, the hotel industry hadn't been disrupted for decades, only to be completely caught off guard by the likes of Airbnb. The internet, combined with near-ubiquitous mobile access, is continually creating very creative entry points for companies to target nonconsumers with more affordable offerings. So I don't believe that the threat of displacement is necessarily greater, but certainly the fact that digital platforms can emerge and expand is something that I just hadn't conceived of early in our research and deserves further study.

Does the rise of "digital transformation" present any anomalies to your theories?

Certainly there are anomalies waiting to be discovered, and further research into digital-focused firms will yield profound insight into the boundaries of disruptive innovation theory. But I believe that the fundamental questions we've been asking for decades now apply just as much in a digital context as they do in an analog one. Who are your best customers? What is your organization capable or incapable of doing? What "jobs" are you trying to help customers get done in their lives? In what circumstances should you integrate, and in what circumstances should you modularize your firm's and product's architecture? Who are the nonconsumers, and what is limiting their access? These strategic questions are universal.

The Theory of Disruptive Innovation predicts what an incumbent will do in the face of a disruptive new entrant. That means incumbents should be well-versed in what not to do. So why haven't more companies solved the innovator's dilemma?

Companies certainly know more about disruption than they did in 1995, but I still speak and write to executives who haven't firmly grasped the implications of the theory. The forces that combine to cause disruption are like gravity—they are constant and are always at work within and around the firm. It takes very skilled and very astute leaders to be navigating disruption on a constant basis, and many managers are increasingly aware of how to do that.

And in my experience, it seems that it's often easier for executives to spot disruptions occurring in someone else's industry rather than their own, where their deep and nuanced knowledge can sometimes distract them from seeing the writing on the wall. That's why theory is so important. The theory predicts what will happen without being clouded by personal opinion. I don't have an opinion on whether a particular company is vulnerable to disruption or not—but the theory does. That's why it's such a powerful tool.

Many of your other theories are vital to understand for companies that not only wish to avoid disruption but also for companies that aspire to be the disrupter. Your Theory of Jobs to Be Done explains how a would-be disrupter nails the right product or offering when an incumbent often can't get it right. Can you explain what this is and why it's so powerful?

My colleagues and I have spent years trying to understand customer behavior—why someone would choose to buy one product or service over other options. What we know is that most companies tend to focus on data to help guide their decisions: They know market share to the *n*th degree, how products are selling in different markets, profit margin across hundreds of different items, and so on. But all this data is focused on customers and the product itself—not what the customer is trying to accomplish in making the purchase. We believe that there's a better way to understand that choice. We call it the Theory of Jobs to Be Done.

There is a simple, but powerful, insight at the core of this theory: Customers don't buy products or services; they pull them into their lives to make progress. We call this progress the "job" they are trying to get done, and in our metaphor we say that customers "hire" products or services to do these jobs. When you understand that concept, the idea of uncovering consumer jobs makes intuitive sense.

Each "job" has not only functional dimensions but emotional and social ones, too. Unless you understand the full context in which your customers are making a choice to "hire" your product or service, you will be unlikely to create the *right* offering for them. You'll just be treading water with them until they "fire" your product and hire one that understands them better. Successful disrupters often nail the Job to Be Done with their offering right out of the gate. Incumbents try to layer more bells and whistles on their product to make it appealing, but in reality, they are missing the fundamental insight of what customers are trying to accomplish. That's why Netflix was so successful in disrupting Blockbuster. Reed Hastings intuitively understood that his customers hired Netflix to relax in their own homes, whenever they wanted. Blockbuster focused on increasing its profitability (for instance, through the horrendous late fees we all sheepishly paid) rather than understanding why we chose to hire a video in the first place. Understanding the Job to Be Done provides a road map for successful innovation.

I know that you relish the opportunity to challenge and strengthen your own theories. There is a sign at your office at Harvard Business School that reads "Anomalies wanted." Are you ever "done" refining your theories?

I have always welcomed challenges to my thinking. I think understanding anomalies—what a theory doesn't explain—helps make the theory better and stronger. We refine the theory with those insights. My own thinking about the theory of disruption has evolved tremendously since I first published its findings in 1995. My goal has never been to be right but to find the right answer. They're very different things. I've long believed that asking the right questions is the only way to get to the right answer. And understanding what questions to ask takes real work.

What do you think people misunderstand about the theory of disruption?

Apart from what you've already mentioned, which is that disruption does not mean "breakthrough" or "new and shiny," far too many people assume that disruption is an event. Rather, disruption is a process. It's intertwined with the resource allocation process in the firm, in the changing needs of customers and potential customers, and in the constant evolution of technology.

There is a growing set of companies that seem to be more fluid in how they approach strategy – like Amazon, Alibaba, and Tencent. Are these companies inoculated against the innovator's dilemma?

This is a very interesting question. I am always wary when we hear that whatever is the high-flying company of the day has solved such a deep systemic problem. Remember, Sears, Digital Equipment Corp., and Eastman Kodak were all once hailed as paragons of good management, until circumstances changed.

That said, there do seem to be some interesting connection points between the companies you mentioned. They have all built organizations that have put the customers, and their Job to Be Done, at the center. They also have demonstrated the ability to manage emergent strategy well. However, they also have been in the fortunate circumstance where their core businesses have been growing at phenomenal rates, and they have had the presence of the founder to help, to personally get involved in key strategic decisions.

One of my former doctoral students, Howard Yu (who now teaches at IMD), noted how important what he called "CEO deep dives" are to wrestling with common innovation challenges, and all of these companies have had the good fortune to have leaders that are ready, willing, and able to do such deep dives. The question for each is, when growth inevitably slows, and when those founders inevitably move on, have they developed the systems, processes, and culture to keep that fluidity? Or, when circumstances change, will the story end the same way it did for other paragons of good management? We will learn something interesting either way.

Anything you got wrong, in hindsight?

I've gotten my share of things wrong. One of the joys of being a professor is that I am challenged on a daily basis by my students, and I know I've learned as much from them as they have from me over the years.

Perhaps most notably, I initially misread the Apple iPhone. When the iPhone first launched, I suggested that Apple had entered late into an established category with a sustaining strategy, and my research showed the odds of success of that strategy was low. I did not see it as disruptive. But then one of my former students, Horace Dediu, taught me that I had framed the problem incorrectly. I viewed Apple as a late entrant into the mobile phone business, where in Horace's view it was an early mover in the "computer in your pocket" business. Horace was right. And, to its credit, Apple then developed a business model that allowed it be a portable PC better than anyone else. People forget this now, but when the iPhone launched, the only applications you could run on it were those that were created by Apple. Indeed, the company was famously protective of its interdependent, proprietary architecture. To Steve Jobs's credit, he and the team created the App Store and opened the architecture up enough to allow an explosion of useful add-ons.

This example reinforced to me the importance of getting the categories right. When someone tells me they are disruptive, the first question I always ask is, "To what?" This is an important question, because disruption is a relative concept.

What questions are you still eager to answer?

Last year I had a conversation with Marc Andreessen about *The Prosperity Paradox*, and we were discussing the role firms play in economic growth. Having just come back from an Airbnb board meeting, Marc described how Airbnb gives ordinary people a platform to offer their services, whether they are cooking a meal for their guests, hosting a class, or giving a tour of their hometown. These citizens would otherwise be unable to participate in the tourism industry, but because of the digital platform of Airbnb, they now can.

It occurred to me that in nearly every case, the firms we profiled to demonstrate how economies are built were those that built physical products. This meant they manufactured, distributed, sold, serviced, and designed goods for a nonconsuming population, resulting in tremendous growth for their firm and their nation. But Airbnb and others like it don't have to do any of those things, and yet they are creating opportunities all over the world. I am eager to explore further the growth potential of digital-first firms and understand what growth looks like in the years ahead.

One of the topics I've loved exploring with you over the years has nothing to do with technology but something far more important, in my mind. I know you've thought a lot about educating children – both in your personal life and in your research. What advice would you give parents of young children about how best to educate their children in today's tumultuous world?

One of my favorite quotes says to let people "be anxiously engaged in a good cause." Far too often, parents smother their children with lists, extracurriculars, and other "good" things, so that children don't learn how to self-manage and regulate their own lives. In our world, that's a vital skill kids need to have because of how distracted we are becoming.

Your theories have provided guidance not only for the senior statesmen of Silicon Valley but for a new generation of entrepreneurs all around the world. And you may have reached a pop-culture pinnacle when you were the answer to a Jeopardy! question a few years ago. But what is it that you would most like to be remembered for?

I want to be remembered for my faith in God and my belief that he wants all of mankind to be successful. The only way to make this happen is to help individual people become better people, and innovation is the key to unlocking ever-more opportunities to do that.

II

The Human Element

2

How Leaders Delude Themselves about Disruption

Scott D. Anthony and Michael Putz

Ever since the 1997 publication of *The Innovator's Dilemma*, researchers, management experts, and businesspeople have discussed, dissected, and debated Clayton M. Christensen's Theory of Disruptive Innovation. By now, the arc of disruption is well established: We know how disrupters enter the market, and we know how incumbents typically bungle their responses to such seemingly insignificant competition. Numerous books and articles have offered to solve the dilemma of disruption, including Christensen's own *The Innovator's Solution* (a 2003 book coauthored with Michael Raynor), which suggests that leaders who understand how disruption transpires can inoculate themselves against the threats and seize the opportunities.

Yet, despite so much insight and advice, the dilemma persists: 63% of companies are currently experiencing disruption, and 44% are highly susceptible to it, according to research by Accenture.[1] And in a thorough analysis of more than 1,500 publicly listed companies, growth strategy consultancy Innosight found that only 52 of them, about 3% of the sample set, had made material progress in strategically transforming their organizations.[2] The default positions, it seems, are to squeeze extra points from profit margins, search for companies to acquire, or simply pay lip service to innovation by setting up token incubators or having executives wear jeans and the occasional hoodie.

Why are companies still so vulnerable to disruptive threats? Our view is that it isn't about not having the right playbook. The problem is that well-intentioned leaders often delude themselves by downplaying disruptive threats or overestimating the difficulty of response. In simple terms, leaders lie to themselves. This means that dealing with disruption is not just an innovation challenge; it is a leadership challenge. This chapter explains these delusions about disruption and offers ways to help leaders avoid self-sabotage.

Cautionary Tales Persist

"Christensen and Raynor have done a superb job of creating a framework for helping to understand industry dynamics and for planning your own growth alternatives." This quote appeared on the back jacket of *The Innovator's Solution*, attributed to Pekka Ala-Pietilä, then president of Nokia. The Finnish company had much to be proud of back then. It was on the brink of taking over the booming cell phone market. Over the next few years, the company would grow into a seemingly unstoppable force. Its stock price surged. Then, in November 2007, *Forbes* ran a prophetic cover with the headline, "Nokia: One Billion Customers—Can Anyone Catch the Cell Phone King?"

Well, yes.

Despite having dominant market share, despite having the resources and capabilities to transition to the smartphone era, and despite having a leader who endorsed and presumably understood Christensen's groundbreaking theories on disruption

It is easy to be in the middle of a disruptive storm and take comfort in data suggesting everything is fine. This is because data lags disruptive change.

(though Ala-Pietilä, admittedly, left the company in 2005), Nokia stumbled. Apple famously entered the market in mid-2007. Google formed the Open Handset Alliance, powered by the Android operating system, later that year. Nokia shares began to slide. In 2013, CEO Stephen Elop sold Nokia's ailing cell phone business to Microsoft for roughly $7 billion. A year later, Microsoft took a roughly $7 billion write-down on the transaction, suggesting the business was worthless.

Nokia's fall is just one of many cautionary tales: Eastman Kodak, Blockbuster, and Toys R Us were all destroyed by disruption. Some of today's great companies look to be similarly snared in their own innovator's dilemma. FedEx started in classic disruptive fashion but today is under threat from Amazon, which could hollow out a significant portion of FedEx's core business by picking off high-value lines between hubs, leaving FedEx (and UPS) with small, unprofitable routes. Or consider Netflix. The disruption poster child could be disrupted itself by Amazon, Apple, AT&T, or Disney. Netflix lacks the diverse portfolio these businesses have, and it may be overly focused on mainstream customers while ignoring the needs of less profitable ones, like all of those young people who prefer creating and sharing short-form videos on platforms such as YouTube and TikTok instead of watching shows like *The Crown*. New habit formation is often an early warning sign of disruptive change. For all its innovation prowess, why hasn't Netflix visibly pursued growth opportunities beyond video streaming and long-form content creation?

And why, after so many years since Christensen presented his original theory and so many cautionary tales, do leaders continue to miss the warning signs? Our view is that the disruptive playbook's leadership section is incomplete. Leaders must learn how to build the individual and organizational capability to confront powerful self-deceptions that inhibit successfully dealing with disruption.

Four Lies Leaders Tell Themselves

Powerful deceptions hinder a leader's ability to respond to disruptive threats and seize disruptive opportunities. Christensen's original research highlighted one such deception, noting how, ironically, listening to your best customers drives the innovator's dilemma. Companies typically focus on satisfying their best customers (usually their most profitable ones) by providing better versions of current solutions while ignoring their worst customers, the ones most likely to flock to cheaper or more convenient disruptive solutions. Of course, that deception is now well known. But other, less obvious lies that leaders tell themselves play a critical role in determining a company's long-term fate. Let's explore four of them.

Lie No. 1: "We're safe." It is easy to be in the middle of a disruptive storm and take comfort in data suggesting everything is fine. This is because data lags disruptive change. CEOs look at their dashboards and think they are OK, but they forget that they are looking in a rearview mirror. BlackBerry is a good example. On April 1, 2008, its co-CEO, Jim Balsillie, gave an astonishing interview on a Canadian chat show.[3] He dismissed the iPhone, didn't mention Android, and smugly said, "I don't look up too much or down too much. The great fun is doing what you do every day. I'm sort of a poster child for not sort of doing anything but what we do every day. . . . We're a very poorly diversified portfolio. It either goes to the moon or crashes to the earth."

It's easy in hindsight to laugh at the quote and especially at Balsillie's hubris. But consider BlackBerry's performance at the time and over the next few years that followed. When Balsillie gave the interview, BlackBerry (then called Research in Motion) had just reported revenues double those in the previous fiscal year, to roughly $6 billion. Over the next three years, revenues tripled, reaching a peak of close to $20 billion. Then, of course, came the crash, and now BlackBerry's revenues are below $1 billion. *Today's data reflects yesterday's reality.*

Ironically, leaders can be adept at spotting disruption in other industries and yet be blind to what's going on right in front of their eyes. Years ago, Innosight and Harvard Business School (HBS) held an event to discuss disruption with organizations that included Hallmark, Intel, Kodak, and the US Department of Defense. Participants were given 20-page case studies highlighting potentially disruptive developments related to their industries. Kodak's case focused on digital imaging, for example, while Hallmark's focused on online greeting cards. But by and large, the cases depicted similar disruptive scenarios. "I thought this was going to be the most boring event ever," admits Clark Gilbert, an HBS professor at the time and now president of BYU–Pathway Worldwide. "We basically had written five versions of the same case." But something surprising transpired. While discussing the cases, it quickly became clear that while executives easily saw disruption in other industries, they missed the forces at work in their own. "It was remarkable," says Gilbert. "None of the companies could see their own problem."[4]

When company leaders finally see the problem, it is usually too late. Leaders must have the "courage to choose" *before* they face the proverbial burning platform. Once the platform is on fire, choices substantially narrow. Mark Bertolini of Aetna provides a good example of a leader who didn't wait that long to act. When he became CEO in late 2010, there was no burning platform. The health insurer had just reported record revenues and record earnings. It would have been easy for Bertolini to execute yesterday's strategy for five years and ride off into the sunset. Instead, Bertolini made the courageous decision to dramatically reconfigure the business, which ultimately resulted in Aetna's game-changing merger with CVS Health in early 2018, a first-of-its-kind combination of retail pharmacy and insurance capabilities. The combination creates the potential for more affordable access to urgent and primary care. The increasing collaborations could break down the traditional health care silos—payer, provider, pharmacy, medical devices—and may signal the beginning of broader health care disruption.

Lie No. 2: "It's too risky." There is a perception that making bold investments in innovation carries systemic risks and that it is safer to stay the course. Consider a large European industrial company Innosight advised. The company was seeking to set a bold new direction and achieve ambitious performance targets. Its leadership identified an opportunity to drive step-change growth by, for the first time in its history, bypassing traditional distributors to deliver highly customized products directly to end consumers. The strategy would require substantial commitment, but it also promised substantial returns, including the ability to spur growth, combat commoditization, and increase margins. Despite consensus and a serious commitment among the executive leadership team to the strategy, the outgoing and incoming board chairs decided—in the bathroom during a break—to significantly reduce the scope of the ambition, perceiving that leveraging digital technologies and bypassing traditional distributors was too risky and not in the company's short- and medium-term interests. The ambitious plan was scrapped after that bathroom break. Both the CEO and CFO departed soon thereafter, leaving a demoralized management team with no clear view for the future except the status quo.

The fear of messing up what has been a proven strategy is powerful, but the reality is that making bold investments in innovation *doesn't* carry systemic risks. New Coke, Apple's Newton, Microsoft's Zune music player, Amazon's Fire Phone, and Google's augmented reality glasses are all examples of big companies that made big bets that led to big write-offs. But while none of these failures was *good*, of course, none sank their companies, either.

"Big moves look like they are really risky. By and large, they are not," declared a Fortune 500 CEO at a 2019 Innosight event. "Because what you lose when you invest a ton of money is the money you invested. It is capped. When you win, you usually create not only an annuity but a new ecosystem that gives you the opportunity to win in new areas."

What carries systemic risk is not making bold investments in innovation. In the face of disruptive change, company leaders consistently invest too little, too slowly, in doing something different. *Companies increase risk by not taking risk*. Walmart executives knew for years they had to embrace online retail, and yet from 1998 to 2015, the company kept redirecting or scuttling significant investments to compete with Amazon and other e-retailers. It spun off Walmart.com in 2000 and brought it back in July 2001. It didn't allow third-party sellers until 2009 and then for years had just six sellers on its website. It invested in an e-commerce site in China in 2011, took a 100% share in 2015, and sold it off in 2016. Its acquisition of Jet.com in 2016 showed promise, but Walmart should have bet boldly years earlier when the capital investment required would have been much lower and the tolerance for losses (necessary to catch Amazon) higher.[5]

Ironically, such late-breaking attempts to catch up on innovation after being too slow are often, wrongly, seen as proof that they offer systemic risk—Walmart is "betting the store" on its web strategy. But that wrongly characterizes what is actually happening. Walmart was very, very late to the party and is now innovating to control the damage. Late-breaking catch-up innovation with a burning (or smoking) platform is not the same as making bold bets early on.

Lie No. 3: "My shareholders won't let me." This deception shows up in various guises. It might be "The activists will pounce on me," or "My shareholders won't like it," or "The market just cares about short-term results." This type of lie—more like an excuse—leads to self-harm. A compelling stream of research by McKinsey shows that companies that take a long-term perspective outperform those that don't. So, paradoxically, those that focus on short-term returns generate lower short-term returns.[6] Indeed, many leaders hide behind the "maximize shareholder value" maxim without understanding exactly what it means. As Michael Mauboussin and Alfred Rappaport, prominent financial experts and coauthors of *Expectations Investing,* once noted:

> Maximizing shareholder value means focusing on cash flow, not earnings. It means managing for the long term, not for the short term. And it means that managers must take risk into account as they evaluate choices. Executives who manage for value allocate corporate resources with the aim of maximizing the present value of risk-adjusted, long-term cash flows. They recognize that to create value, a company must generate a return on its invested capital that covers all of its costs over time, including the cost of capital. These executives are not fixated on the short-term stock price but rather on building enduring long-term value that ultimately shows up in a higher stock price.[7]

Further, there are clear examples demonstrating that you can actively sell your shareholders on a new story, as long as it is indeed a convincing story and you demonstrate early success. But that doesn't make it easy. Aetna's Bertolini recalled a tense meeting early on in the company's strategic transformation where he fielded questions from investment banking analysts: "I walked into a room of analysts and I said, 'You either think of me as stupid or that I'm lying to you, neither of which makes me want to spend more time with you.' I have had shareholders who have said to me, 'Why don't you double your dividends?' Well, I want to invest in the company. So I said that one of my largest shareholders should get the hell out of my stock."[8] In 2018, *Harvard Business Review* identified Bertolini as one of the 50 best-performing CEOs in the world and the highest-performing CEO in the health insurance industry.[9] Bertolini's decision to stand firm in the face of criticism helped drive Aetna's successful transformation.

Lie No. 4: "My people aren't up to the task." Leaders often use their own people as an excuse not to take action. It's a convenient lie that puts the burden of inaction on others—or worse, requires leaders to take dramatic action that may not be required. For example, in 2018, Biogen CEO Michel Vounatsos told a group of CEOs that transforming his company had required turning over a staggering 80% of his top leadership team. "People resist change," he said. "You need to find the leaders in the room who will be the ambassadors to the future."[10] Vounatsos's strategy underlines a popular perception that changing a company requires changing the staff. It is too early to tell whether Vounatsos's method will pay off, but there is an obvious risk of destroying a significant amount of institutional knowledge. The difficulty and pain of the "rip and replace" strategy is perhaps one reason why only 3% of companies researched by Innosight were embarking on significant strategic transformations.

As a counterexample, DBS, the largest bank in Southeast Asia, went from a stodgy regulated bank to an innovative digital powerhouse without dramatically changing its workforce. Soon after becoming CEO in 2009, Piyush Gupta set a challenge: Given increasing technological change, DBS had to function like a 27,000-person startup. That meant embracing new behaviors, such as agile development, customer obsession, and experimentation. DBS's culture change effort rested on the fundamental belief that its staff had the inherent capabilities to become innovators but lacked the tools and support to realize them. Under the leadership of Paul Cobban, chief data and transformation officer, the bank redesigned its offices, changed its meeting styles to encourage equal participation and greater collaboration, and introduced a new innovation team with an unusual mandate: *Under absolutely no circumstances should the innovation team innovate.* Instead, the team's mission is to enable the broader community to incorporate the behaviors that enable successful innovation.[11]

Transform Thyself

To see these lies for what they are and successfully transform their organizations, leaders first need to transform themselves. Successfully responding to disruption requires executives to simultaneously reinvent today's business while creating tomorrow's business. More specifically, they have to find new ways to solve customer problems while at the same time scoping out new growth opportunities. The challenge isn't just that these missions are in conflict and involve periods of chaos and uncertainty; it also is that they require fundamentally different mindsets and approaches.

To see these lies for what they are and successfully transform their organizations, leaders first need to transform themselves.

Research by longtime Harvard professor Robert Kegan found that most leaders lack the cognitive flexibility required to "toggle" between being disciplined and entrepreneurial. Kegan terms this flexibility *self-transforming*, where leaders have the ability to "step back from and reflect on the limits of our own ideology or personal authority; see that any one system or self-organization is in some way partial or incomplete; be friendlier toward contradiction and opposites; [and] seek to hold on to multiple systems rather than project all except one onto the other."[12] Unfortunately, other research suggests that no more than 5% of high-performing managers have achieved this level of leadership.[13]

It's not surprising that so many leaders lack this capability. Most grew up in a world that was either disciplined or entrepreneurial but rarely both and almost never both at the same time. And leadership development (with rare exceptions) hasn't caught up with this emerging need. To transform themselves, leaders must focus more on mindsets, awareness, and inner capacities to combat basic biases that make it hard to make decisions in uncertainty and toggle between different frames. There are no quick fixes here. But research increasingly suggests the best starting point is to embrace what broadly goes under the term *mindfulness*.

To some, the word might sound squishy and New Agey, but meditation and related practices that use breathing to tune into thoughts and sensations have widely documented health benefits, such as boosting energy and lowering stress. More critical, and for our purposes here, mindfulness boosts awareness, increasing a person's ability to step back, pause, and become aware of not just habitual thought patterns but also emotional reactions. As Potential Project managing director Rasmus Hougaard has noted, mindfulness is not about just doing more but also seeing more clearly what is the right thing to do and what is just a distraction.[14]

Mindful leaders can, for example, "see" their reactivity, giving them the tools to identify and overcome the deceptions of disruption. A mindful leader is better at toggling between different mindsets: a disciplined focus on transforming today's business and more entrepreneurial thinking to create tomorrow's business. Mindfulness is a powerful, scientifically validated tool for improving self-awareness, which is a critical and underappreciated tool for senior leaders confronting the challenges of disruptive change.

Some leaders who have successfully managed transformative change have touted the value of mindfulness. Aetna's Bertolini was an early advocate of advancing meditation programs at his company, and in 2014, the company hired a chief mindfulness officer. Bertolini credits mindfulness for easing chronic pain he suffered after a skiing accident and when recovering from a rare form of cancer. He says it also improves his ability to process information and make sharp strategic decisions: "With so many things going on, whether in a small or large organization, you can get frozen by attempting to process it all instead of being present, listening, and focusing on what really matters."[15]

Another example of the power of mindfulness comes from Pierre Wack, who advanced and popularized the idea of scenario planning while working at Royal Dutch Shell in the 1970s and 1980s. Wack was influenced by Russian guru Georges Gurdjieff and practiced meditation in India. Successful scenario planning, Wack noted, requires "being in the right state of focus to put your finger unerringly on the key facts or insights that unlock or open understanding."[16] He noted that the value of scenario planning is not about developing specific plans that will actually be implemented or getting to the "right" scenario but about helping leadership understand that the future can be dramatically different from the present, while fostering a deeper understanding of the forces driving potential changes and uncertainties. The approach, he said, gives managers something "very precious: the ability to reperceive reality."[17] By sharpening his ability to toggle between present reality and future possibility, Wack and his team transformed scenario planning from a passive manipulation of data into an active tool to stretch thinking and advance discussions. This helped Shell to see what others missed and weather oil shocks in the 1970s and 1980s significantly better than its competitors.

Transform Thy Organization

Of course, it is not enough to transform just the person at the top. Too often you see a *single-shot transformational leader*. That is, a leader seems to transform an organization, but then the organization backslides when the leader departs. For example, A.G. Lafley drove substantial change at Procter & Gamble as CEO from 2000 to 2009, but the company stumbled so badly when he stepped down that he was asked to return. Lou Gerstner was an icon for transformation during his time at IBM, but by the time Sam Palmisano turned the CEO reins over to Ginni Rometty, IBM had missed the cloud computing revolution and its touted Watson platform was struggling to deliver results commensurate with its hype. Tim Cook has been a strong steward at Apple, boosting growth and strengthening its services business, but he simply hasn't matched the disruptive magic of Steve Jobs. Single-shot leaders might have the *personal* ability to toggle between different mindsets, but they seem to struggle to codify core elements of their unique approach and institutionalize them.

Kegan and coauthor Lisa Laskow Lahey noted in *An Everyone Culture* that you can create a *deliberately developmental organization* (DDO) that consciously upgrades an entire organization's capacity to grapple with disruption. Bridgewater, the world's largest hedge fund, is a good example. It seeks to base the organization not on founder Ray Dalio's charisma or his intuition but rather on decision rules, which Dalio calls *principles*, hardwired into its systems. Some of the fundamental principles include radical transparency, where the goal is to review people "accurately, not kindly"; recognizing that internal exploration and struggle is important ("pain + reflection = progress"); and sharing and supporting project work with complete transparency. Bridgewater gives employee feedback not just to boost short-term performance but to enhance long-term capacity. It consciously helps its employees develop reflective muscles to understand defensive routines and blind spots and to improve their ability to acquire, process, and make sense of multiple forms of data. This commitment to developing everyone's "sense-making" capacity as a mission-critical component of long-term performance sets DDOs like Bridgewater apart.

Two final examples worth mentioning for their transformative efforts are SAP and Johnson & Johnson, which are helping staff develop creatively, emotionally, and mentally to tackle larger challenges such as disruption. SAP has trained more than 10,000 of its employees to use meditation to improve self-perception, regulate emotions, and increase resilience and empathy. Participants report double-digit increases in their personal sense of meaning, their ability to focus, their level of mental clarity, and their creative abilities.[18] For its part, Johnson & Johnson has long focused on employee well-being. Recent efforts center on energy and performance, with a stated goal to help employees achieve "full engagement in work and life." Participants answer diagnostic questions such as, "Are you present in the moment, focused, and fully aware?" and peer reviewers assess whether "their self-image is keeping them from being the person they wish to be." Leaders serve as active role models for building these capabilities. CEO Alex Gorsky, for example, has long worn a fitness tracker and speaks publicly about the link between mental well-being and productivity.[19]

In the first two decades of the 21st century, scholars and practitioners fine-tuned the *technical* solution to the dilemma of disruption. It is high time for that community to more fully define a *human* solution—one that starts with senior leadership and carries through an entire organization. If you want to defeat the dilemma of disruption, you must start by defeating delusions about disruption. That challenge starts at the top.

3

Overcoming the Innovator's Paradox

Jeff Dyer, Nathan Furr, and Mike Hendron

Having a great idea is essential to innovation, but that's only half of what's needed. Securing the resources to implement the idea is just as important—and potentially more difficult. The inventor Nikola Tesla, for example, came up with several transformative ideas—for electric induction motors, wireless telegraphy, radios, and remote control—but he died penniless because he couldn't line up the resources to commercialize them. In contrast, Thomas Edison, arguably less brilliant, died wealthy and famous because he was good at both coming up with ideas and winning the necessary support to turn them into reality.

Every innovator faces what we call the *innovator's paradox*.[1] Quite simply, the more novel, radical, or risky the idea, the bigger the challenge in acquiring the necessary resources. Although many people say they like radical ideas, the greater the risk and uncertainty, the more skittish would-be supporters (investors, bosses, partners, and so on) become. Many great ideas die on the drawing room floor because entrepreneurs fail to persuade others of their potential. Before jumping in, potential backers want ideas to be proven or the uncertainty reduced in some meaningful way. The ability to overcome the innovator's paradox is key to becoming a successful innovator, whether you work inside a company or are trying to launch a new venture.

How can you prove an idea can work if you lack the resources to start developing it? Even the leanest of lean startups needs a basic level of resources to test its ideas, and truly big ideas need more. In a world where more ideas than ever are competing for resources—whether that's financial backing, good team members, permission to pursue an idea, or customers—innovators who learn how to win support are the ones who gain traction.

To understand how successful innovators and entrepreneurs go about winning critical support, we conducted more than 100 interviews. We spoke with well-known innovative leaders such as Jeff Bezos (Amazon), Elon Musk (Tesla and SpaceX), Marc Benioff (Salesforce.com), Shantanu Narayen (Adobe), Indra Nooyi (PepsiCo), Mark Parker (Nike), and Jeff Weiner (LinkedIn), as well as less well-known innovators. The good news is that the capacity to win support for your ideas isn't something you have to be born with. Indeed, by leveraging the tools and practices we describe in this article, you can develop this skill. As Benioff, founder and co-CEO of Salesforce, told us, "My ability to generate innovations [for Salesforce] has basically built up over time." Over 20 years, he has accumulated what he calls *innovation capital*, which has allowed him "to try new things—to change the organization, change the products, change what needs to be changed." Benioff claims that as his ability to "try new things" has grown, his associates have built their own innovation capital that helps them sell their own ideas.

Politicians with political capital can get others to join them in pursuing their objectives; business leaders with innovation capital can attract the resources needed for innovation to flourish. Innovation capital consists of four factors: who you are (innovation-specific human capital: your capacity for forward thinking, creative problem-solving, and persuasion); who you know (innovation-specific social capital: your social connections with people who have valuable resources for innovation); what you have done (innovation-specific reputation capital: your track record and reputation for innovation); and the things you do to generate attention and credibility for yourself and your ideas (what we call *impression amplifiers*).[2]

While all four elements contribute to one's innovation capital, the first three are built over time through sustained effort. All of them are important, but in this article we will focus on the fourth element: the actions you can take now to win support for your ideas.

The Research

- To examine the techniques companies use to win support for new ideas, the authors developed more than 50 case studies and conducted more than 100 interviews with innovators.
- They reviewed academic research in a variety of fields related to innovation, including human capital, social networks, reputation, decision-making, cognition, communication, and persuasion.
- In addition, they conducted a study on the relationship between innovation reputation and market value among S&P 500 companies.

How Amplifiers Work

On the surface, the impression amplifiers we heard innovators talk about are not particularly surprising. To the contrary, they include the types of things you might expect:

- **Comparing:** Finding the right analogy to convince supporters your idea will succeed.

- **Materializing:** Making an abstract concept tangible, visible, and real.
- **Storytelling:** Crafting a narrative that gives listeners a reason to believe.
- **Signaling:** Connecting to other credible groups that confer legitimacy to your idea.
- **Applying social pressure:** Creating a sense of scarcity (the feeling that people need to act now or they'll miss out).
- **Committing:** Convincing others through a visible, personal, or irreversible action.

Although these impression amplifiers may sound simple, there is an art to using them well. Each one taps into a distinct and powerful psychological principle supported by research. Most of the innovators we studied didn't consciously realize that they used these techniques. They seemed to intuitively recognize that when trying to win support for significant new ventures, they had to go beyond conventional tactics such as data analysis, financial forecasting, and strategic planning. To overcome the innovator's paradox, they often relied on the impression amplifiers described here.

Comparing: "We're the Airbnb of the X industry." In 2000, when Robin Chase cofounded Zipcar, an online service that allowed drivers to rent cars for short periods of time, the obstacles to success were massive. At the time, only 50% of Boston-area residents had internet access, and only 25% had mobile phones. Hardly anyone spoke about platforms or the "sharing economy," and there was little innovation in the transportation sector. Prospective investors had a hard time grasping why people would be interested in an internet-based car-sharing platform. Many of them just listened and nodded politely.

The breakthrough came when Chase discovered how to use the comparing amplifier, drawing an analogy between her idea and something people knew and saw in a positive light. Initially, Chase thought she was using this technique effectively: She described the business as *car sharing*. But that comparison actually turned many people off. (Imagine if you described the hotel business as *bed sharing*.) Chase realized she needed a better comparison. To find it, she carried around note cards and tested names and analogies with people she met. Finally, she latched onto a comparison people understood and *liked*: automated teller machines. ATMs allow individuals to get cash whenever they need it. Zipcar, Chase explained, "was like an ATM, but it's *wheels* when you want them."

Armed with the right analogy, she started to win investor support for her idea. "When I used the comparison 'wheels when you want them,' investors finally got it," she told us. "Words really matter, and getting the comparisons right was nontrivial." In the end, Zipcar was sold to Avis in 2013 for $500 million. But it might never have gotten off the ground without the right comparison.

Comparing turns your idea into something people can relate to and, ideally, get excited about. It draws on the psychology of mental shortcuts (heuristics) and how we use analogies to understand new concepts, form judgments, and make decisions. In practice, research shows that startups that use analogies and metaphors in their prospectuses achieve higher valuations during initial public offerings.[3]

How to use comparing. Our research suggests that effective comparisons follow several key principles:

- The best comparisons capture *both* the opportunity and the solution. The problem with Chase's initial car-sharing analogy was that it only described the solution and didn't excite people about the opportunity. Customers and investors could relate to the ATM analogy and knew how ATMs had revolutionized the banking industry.

- For any given idea, there are often multiple potential benefits. So, when looking for the right comparison, see what resonates with your audience. Zipcar, for example, offered features such as convenient scheduling, booking by the hour, and "green" mobility—but convenience struck the strongest chord.
- Comparisons are especially valuable for complex ideas. If we said that Ethnamed is a Chrome plugin that allows users to bind email addresses with crypto addresses, most people would be puzzled. But if we said it's like Venmo for cryptocurrencies, potential users would more easily understand it and see the potential value.

Materializing: Show – don't just tell. When technology entrepreneur Elon Musk founded SpaceX in 2002, no private rocket company had ever won a contract from NASA. Another rocket company, founded by billionaire Andrew Beal, had recently filed for bankruptcy after failing to persuade NASA to buy its low-cost disposable rockets. So Musk, whose idea was to develop reusable rockets, faced an uphill battle. Leaders at NASA, in Congress, and at large aerospace companies saw SpaceX as a joke—a bizarre effort by a Silicon Valley hotshot that was destined to fizzle. The company struggled to pitch its ideas to key decision makers. "At the beginning, we had to beg NASA to even pay attention," recalled Lawrence Williams, a former SpaceX executive.

How did Musk break through the resistance? By *showing* people what SpaceX planned to do—a tactic we call *materializing*.

On the eve of the centenary celebration of the Wright brothers' first powered flight in 1903, Musk loaded a recently completed Falcon 1 rocket onto a gigantic trailer and drove it down Independence Avenue and along the National Mall in Washington, D.C., before parking it in front of the National Air and Space Museum. He invited the event's attendees to inspect the rocket. The spectacle got NASA's attention—and showed off what SpaceX was capable of building. The following month, NASA administrator Sean O'Keefe dispatched a team to visit SpaceX in Southern California, and they were impressed—so much so that the agency invested $140 million in 70 SpaceX flights over the next decade.[4]

In essence, Musk advanced the idea of space travel by turning it into something tangible—in this case a physical prototype. Because our brains prefer the tangible and visible over the abstract,[5] people respond to visual objects and remember them better.[6] Making abstract ideas tangible also makes them seem more realistic and believable. Research shows that entrepreneurial teams that materialize their ideas are less likely to fail, presumably because they are better at winning the support needed to turn the idea into reality.[7]

How to use materializing. Being skillful at materializing is an art, and it's essential to understand what your supporters will be looking for when you materialize your idea. In some cases, the critical question will be "Can it work?" and in others it will be "Will customers want it?" Materializing along the wrong dimension or in the wrong way can lead to problems—something the founders of The Boring Company, a Los Angeles-based transportation startup, discovered the hard way. The company had developed an ambitious plan to build a high-speed system that would reduce traffic by transporting vehicles and people through vacuum tubes and tunnels. But the founders failed in their attempt to bring their idea to life. Their prototype was badly constructed, leading to test rides that were both slow and bumpy. Transportation officials soon lost confidence in the concept and withdrew their commitments.[8] What can companies do to avoid similar mistakes?

- If the big unknown in your supporters' minds is whether customers want your offering, focus first on materializing the demand as opposed to whether it works. For example, when Rent the Runway founder Jennifer Hyman came up with the idea of renting designer dresses over the internet, it wasn't clear whether women would rent a dress for a special event. So before setting up the logistics and distribution system, Hyman did low-cost experiments, first with physical locations and then online, to see how women would react. The data convinced Bain Capital to invest in the venture.
- If the key question for your supporters is more technical—"Will it work?"—focus your efforts on meeting that requirement. For example, when the executive team at Caterpillar resisted a radical new way to create a hybrid heavy excavator that would use less energy than a traditional diesel excavator, Ken Smith, head of the heavy excavator division at Caterpillar, put the team to work, showing it could be done. Although the resulting prototype was so ugly that they called it Medusa, it worked, and it convinced executive leadership they could create a new excavator using 33% less energy.
- Materializing can be particularly important for radical ideas. The more novel the idea, the more likely that simply asking customers for feedback can mislead you. For example, many customers hated the abstract descriptions for such innovative

products as the Aeron chair, Reebok Pump shoes, and even the iPhone. But when they actually tried these products, they changed their tune.

Storytelling: Triggering emotional reactions. Soon after Vanessa Quigley finished describing her idea for a new photo-sharing service aimed at parents back in the fall of 2013, Gavin Christensen, the managing partner of Salt Lake City-based Kick-start Seed Fund, told his partners that he hated the product but thought they should invest anyway. How did Quigley convince a hard-nosed venture investor to get behind an idea he disliked? More than anything else, she told a good story.

Although many people have heard about the power of stories in business, few have the ability to use them well. In most organizations, stories are chronologies, mission statements, or thinly veiled manipulations. What made Quigley's different was her use of a narrative, with strong characters, conflict, resolution, and, most of all, emotion. She recalled the day when she found her 7-year-old son in tears over a photo scrapbook his kindergarten teacher had made for him. She described the guilt she felt—after all, she had hundreds of photos on her phone that she had never shared with her son or other family members. Based on this experience, she created an app called Just Family to make it easier to share family photos.

Christensen remembers being impressed by Quigley's story but frustrated by the product itself. Each time he tried to use the app, it crashed. Yet when Quigley talked about the underlying emotions and the need to capture and share memories, he was swayed. "The need was so fundamental, and Vanessa made it so personal," he said. "I knew that they would figure out how to solve it eventually." The company, which now operates as

Studies in neuroscience have shown that stories can cause the minds of storytellers and listeners to "sync."

Chatbooks, allows users to create printed scrapbooks with ease from Instagram photos. Within two years, the business had revenues of more than $30 million.

Successful storytelling triggers emotional connections that can prompt listeners to suspend disbelief and even to take action. Studies in neuroscience have shown that stories can cause the minds of storytellers and listeners to "sync" (that is, to develop the same brain patterns) and "couple" (connect and anticipate what the other experiences),[9] an effect that increases your ability to persuade your listeners: Their brains literally start to match yours. Moreover, psychology research underscores how good stories can capture our attention, transport us to the world of the storyteller, and release brain chemicals that increase the likelihood of persuasion and action.[10]

How to use storytelling. Although most leaders think they know how to use stories, there are basic rules many people forget. Effective stories are personal, create a narrative arc, vividly paint the opportunity or threat, and create an emotional connection. Master storytellers use several tactics:

- They develop a narrative arc, with characters, conflict, and resolution (the classic hero's journey). For example, when David Hieatt, a successful advertising executive and entrepreneur, moved to Cardigan, Wales (population: 4,000), he was struck by the fact that the town had once been the center of jeans manufacturing in Britain. When talking about his new apparel company Hiut Denim, Hieatt found it effective to describe the failed jeans industry, the people who were left behind, and his quest to become a force for good by bringing jobs back to the area.

- They tie listeners into the story. Consider Alibaba's Jack Ma. Although Ma had flunked out of school multiple times and failed at his first two startups, he was still able to convince investors to back his idea for Alibaba.com.[11] He was particularly effective at getting both employees and investors engaged with his story by telling them about the company's historic quest to create a better future for small-business owners in China by opening the frontier of the internet for them.
- To not set unrealistic expectations around distant or long-term objectives, storytellers avoid making specific claims and focus on more abstract promises.[12]

Signaling: Building legitimacy through others. Over a decade ago, Brad Jones, an executive at Melbourne, Australia–based ANZ Bank, had an idea for addressing a serious problem affecting low-income people in emerging economies such as Cambodia. Many of these people were "unbanked" and thus were more likely to lose their cash, have it stolen, or pay big commissions to send money over any distance. Jones envisioned a mobile money platform that would allow the "bottom of the pyramid" to access the same benefits as people with bank accounts. Although Jones secured some initial funding for his Wing Cambodia venture from his bank just before the 2008 financial crisis, it faced the chopping block as the bank looked for ways to save money during the impending downturn.

In his efforts to find funding, Jones began reaching out through his social network and was able to convince ANZ Bank's head of corporate social responsibility to feature Wing Cambodia in the bank's annual report. He was also able to obtain a grant from the Australian government to support the initiative, which brought additional attention and legitimacy to the venture and helped insulate it at a time when other projects were being slashed.

Jones used signaling, which involves making connections—such as alliances, relationships, and awards—to build legitimacy and support for Wing Cambodia. Signaling works because of the human tendency to follow others we admire or see as experts. Research examining more than 1,000 startups found that if the startup had the endorsement of a development organization (such as an incubator), its chances of receiving funding increased from 5% to 44%, even after controlling for the ability of the founders.[13]

How to use signaling. The goal of signaling is to prove to supporters that what you're doing is legitimate—and that others believe in you. As you try to create these signals, keep in mind your audience and what signals they will find most valuable:

- Although signaling is often helpful, research suggests that it's particularly valuable in the early stages of a project, when it is difficult to assess the project using more objective measures.[14] Try to obtain endorsements, awards, or notable acknowledgements to increase your chances of winning support.
- Select the right types of endorsers for your project. When the uncertainty is tied to the technology ("Can we build it?"), it's important to get endorsements from individuals or institutions seen as the "experts" (such as leading scientists). On the other hand, when there's uncertainty around demand ("Will people buy it?"), you'll want endorsements that can influence potential customers and create demand.
- Institutional endorsements can be more valuable and enduring than endorsements from individuals because supporters assume that institutions have done more to verify

the legitimacy of your idea and are taking a greater risk by endorsing you. In the case of Wing Cambodia, for example, the institutional endorsements in the form of exposure in the bank's annual report and the Australian government grant were more credible than the word of any one person.

Applying social pressure: "Don't be left out." A junior manager in a large organization was struggling to get his company to accept the use of robots for tracking inventory. He had done a lot of research on robotics and felt that there were now good opportunities to deploy them. But the company's executives weren't sold—most felt the technology was too immature and risky. So the junior manager tried a different approach: He convinced the CEO of one of the company's recent acquisitions—a much smaller company—to test the robots in his stores. When the other, more skeptical, executives heard about the experiment and the positive results it was generating, they immediately agreed to test it themselves.

Creating a fear of missing out (often known as FOMO) can help spur commitment from those who worry that if they don't act quickly, they might lose an opportunity. It works on psychological principles tied to scarcity: People want things other people want, and they are drawn to items that are either scarce or difficult to obtain. In a famous study, researchers found that customers wanted a product more if they were told it had sold out.[15] In the example above, the executives feared that other companies were moving ahead with the robotics technology, and they didn't want to be seen as laggards.

How to leverage social pressure. Communicating a sense that time is of the essence can get reluctant supporters to move in the right direction, but you need to do this carefully. People skilled at using this impression amplifier apply the following tactics:

- Fear of missing out works best in competitive situations. Having discussions with multiple potential sponsors, employees, or partners in parallel (rather than sequentially) and letting them know that others are considering the opportunity can be very effective.
- Create the right circle of competition among backers. The purpose of using social pressure is to overcome potential supporters' tendency to hold off on making a commitment until you've resolved all the uncertainties of the project. Research shows that having investors with profiles and orientations that are too similar can be a problem.[16] You can overcome this weakness by going after sponsors who are sufficiently different from one another and don't know how to predict one another's behavior.
- Prep candidates in advance. Few people are willing to commit right away. It's helpful to expose potential partners and investors to an idea multiple times before trying to get them to sign on. This approach has been shown to work well for many types of entrepreneurs.[17]
- Don't overplay your hand. Using fear of missing out should not be about creating false expectations or misrepresentation—that will backfire. Rather, it's about using a competitive dynamic to get people to back your project.

Committing: Putting skin in the game. Denver Lough, a third-year resident at the Johns Hopkins University School of Medicine, faced a dilemma. Before beginning his training in facial plastic and reconstructive surgery, he developed an idea for generating new skin using a patient's own cells. He had done some preliminary lab tests showing that the technology worked on mice and felt confident that the same approach would work on

Fear of missing out works best in competitive situations.

humans. If he took the idea to a major medical center, Lough thought he could get lab space and resources to test the idea. But the medical center would want a large equity position, and it wouldn't be able to provide the resources needed to launch a company to commercialize the technology. Lough took his idea to a number of venture capitalists and angel investors. But the investors he contacted wanted too much control or weren't willing to invest in the technology without human clinical trials and patents protecting it from imitators. Lough needed money to develop and prosecute patent applications, and to prove the technology.

Lough left Johns Hopkins early, forgoing future opportunities to earn in excess of $750,000 per year. Instead, he and a fellow resident, Ned Swanson, founded PolarityTE, a biotech startup that would develop Lough's technology for regenerating new tissue. The fact that the two young researchers were willing to take the risk proved critical to their ability to attract investors and key talent. As one investor noted, "They were willing to get up and leave and lay it all on the line, even when they were so close to finishing that they could see the light at the end of the tunnel. That said a lot."

Committing involves standing up for an idea or project you believe in and showing other potential supporters that you are willing to put skin in the game. This might include investing your own money or giving up an attractive employment option. Making a big—perhaps even irreversible—commitment to a project gets people's attention and boosts the likelihood that others might become backers as well. Committing is particularly effective when the innovation is complex or when details cannot be fully revealed because the innovator wants to protect trade secrets. The psychological principle of loss aversion (when

It’s assumed that those willing to take major risks are confident they’ll be successful, perhaps because they know something we don’t know.

the desire to avoid losses has a stronger pull than the opportunity to achieve the equivalent gains) suggests that we don't expect people to do things that will lead to big losses.[18] Indeed, it's assumed that those willing to take major risks are confident they'll be successful, perhaps because they know something we don't know.

How to use committing. Using committing well requires some care. We don't typically recommend that you bet the farm (after all, what happens if things don't work out?). So here are some tactics to show people you are confident without risking everything:

- Create credible commitments, ideally with your abundance rather than your scarcity. For example, in the early days of Tesla, investors were skittish about investing in Musk's bold and expensive plan. However, when Musk offered to fund the company with his own capital, others joined in, believing that his willingness to bet his own money meant he was confident the business would succeed. (We should note that Musk had plenty of capital to work with; unlike many other innovators, it's doubtful he was putting his family or his future in danger.)
- Have a plan B. Although commitment involves sending a strong signal, you shouldn't paint yourself into a corner, in case things go poorly. In scratching below the surface of seemingly "irreversible" commitments, we found that innovators often had a plan B. Lough, for example, wouldn't have been able to return to his residency at Johns Hopkins. But he knew his skills would have value at other medical institutions or biopharma companies if the venture didn't work out.

The successful innovators we studied, including those who have built highly successful organizations, started with a creative idea. But they quickly learned that creativity was not enough. "Ideas are easy in many ways, and there's no shortage of great ideas," Nike CEO Mark Parker told us. "It's the ability to bring those ideas to life and at scale that becomes important." Parker understood that successful innovation requires resources, and persuading others to commit those resources is no small task.

To promote their innovations, effective leaders rely on a set of tactics to gain attention and credibility for themselves and their ideas. Successful entrepreneurs and innovators everywhere face the innovator's paradox: They need to convince others to back risky ideas that may ultimately flame out. Although all of us can build up human, social, and reputation capital over time to win support for our ideas, in the short term we must learn to apply other techniques to shore up our innovation capital so that we can assemble resources that can turn creative ideas into reality.

4

A Crisis of Ethics in Technology Innovation

Max Wessel and Nicole Helmer

Cambridge Analytica has become a household name, synonymous with invasion of privacy. Its controversial entanglement with Facebook was a wake-up call about how we share information online. Of course, Cambridge Analytica is gone now, and Mark Zuckerberg has survived so far. But the fallout for Facebook feels never-ending: the initial stock drop, the congressional testimony, a record-breaking $5 billion fine from the Federal Trade Commission, a class-action suit approved by a federal judge,[1] and another uncomfortable grilling in Congress.

The Facebook scandal is a cautionary tale for executives and consumers alike. But the lesson is much bigger than one about so-called fake news. The hasty reconstruction of value chains around new technologies is introducing and exacerbating ethical concerns across industries. It's a free-for-all race as companies compete to impress users with new capabilities, and what's at stake isn't just which ones survive but whether we are able to sustain a civilized society or end up in a high-tech Wild West.

Facebook ushered in a new era of publishing by building the world's largest content creation and distribution network, amassing billions of users. It invited content makers and advertisers to subsidize those users on a platform that many people feel they can't live without. No longer was the media value chain being orchestrated by a few large organizations; Facebook was opening up markets by enabling anyone with a keyboard and an internet connection to effortlessly plug into the world's largest distribution system. In effect, Facebook broke apart the media value chain and simultaneously re-created it around the company's application programming interfaces (APIs).

But as Facebook helped transform an industry ecosystem, it didn't concern itself with editorial ethics. It sold access to its user base—to companies like Cambridge Analytica—while maintaining distance from anything posted on its own platform. Content creators could tap into end-user data to precisely target their messaging, whether the information they were putting out was false, misleading, or true. Driven by demand from billions of users, Facebook focused only on ensuring that the content on its network amassed clicks.

In this new world of publishing—where authors, editors, and distributors are separate entities pursuing their own interests—the scandalous consequences may seem predictable. After all, accountability also splinters with the rest of the value chain. But when no one steps up to maintain ethical standards across the system, we all suffer in the end.

Facebook is just one example of the evolving—and murky—world of self-defining ethics in technology. In this chapter, we argue that as technological systems rapidly restructure, ethical dilemmas will become more common and that well-understood theories can help us predict when and where problems may arise. Executives across industries find it enticing to democratize access to cumbersome markets like health care, lending, and publishing. But if you're the executive who happens to decouple consumer protection from mortgage lending, all the positive intentions in the world won't protect you from the unavoidable backlash.

Bottom line: Predicting where your industry will stumble within this new world can make the difference in ensuring your business flourishes with its reputation intact.

Modularizing Faster Than Ever

To be clear, this is not about a few software bugs resulting from a "move fast and break things" mentality. This is about leaders, acting in the best interest of markets and consumers, enabled by the ubiquity of the internet, who unintentionally sidestep the ethical protections that underpin society as we know it. To understand the imminent ethical crisis and why current circumstances are so different, we need to understand how value chains emerge and why even responsible technology companies may overlook their ethical obligations.

In 2001, Clayton M. Christensen, Michael Raynor, and Matthew Verlinden published a lauded article in *Harvard Business Review*, "Skate to Where the Money Will Be."[2] It explained what they called the Theory of Interdependence and Modularity. The theory holds that when new technologies emerge, they tend to be tightly integrated in their design because dependence among components exists across the entire system. To combat this fragility, one entity must take tight control of the system's overall design to ensure performance.

Consider the early iPhone. Apple controlled the software, hardware, and even the network—to give users the best experience. There was one size, one browser, and one carrier. Features were eliminated to support battery life, capacitive touch, and call quality. In Christensen's language, the design's *interdependence* was critical, as the phone itself struggled with basic performance issues related to its core function of voice communication. Only Apple's unequivocal control made the product reasonably competitive.

Christensen and coauthors argued that, over time, the connections among different parts of complex systems become well understood. Each element's role is defined. Standards are developed. To use Christensen's term, the industry becomes *modular*, and an array of companies can optimize and commercialize small, specific components with no meaningful impact on overall system performance. Today's iPhone consumers can choose their screen size and phone thickness, the App Store is filled with tools and games from millions of different developers, and phones are available on any network. An entire smartphone industry now exists whereby consumers can pick and choose practically everything about their phones, and the software on them, to meet their individual needs.

For any new technology industry, modularization is the end state; it benefits consumers and grows the pie. Since one company no longer needs to take responsibility for the entire system, every company is free to focus on whichever elements they deem to be strategically advantageous. Christensen, Raynor, and Verlinden counseled companies to anticipate how their markets would become modular and to compete in the places most difficult to master. In the smartphone arena, chipset makers and mobile app companies gobble up all the profit in the system as they tackle the most differentiated parts of it. Playing off a famous hockey tip from Wayne Gretzky in their *HBR* article title, the authors coached strategists to head to "where the money will be," not where it is today.

But modularization is a double-edged sword: The disaggregation of development responsibility also means the diffusion of responsibility for ethical outcomes.

And today's reality is that modularization is accelerating across industries. The internet standardized communication, architecture, and information exchange in every function, allowing new businesses to turn a profit by perfecting ever-more-narrow slices of a value chain.

Consider Lyft. When the company went public in March 2019, its filings recognized the risk that it relied on critical third parties for payments, financing, web infrastructure, background checks, and other significant technology components. It is a massively successful business, but many of its core processes are delivered through the combined services of other vendors. We'd expect similar risks to be identified in the filings of almost every outgoing IPO.

The rise of companies focused on simple components of complex systems has created a virtual à la carte menu from which would-be disrupters can tailor new, complex products according to customer demands. The result: a virtuous cycle that has caused a whirlwind reconstruction of value chains in every industry.

In our increasingly modular world, companies can quickly tailor products to user demands; innovation and opportunity flourish, but so do the potential risks—not just to a company's bottom line and reputation but also to society at large. Innovation might be able to move with lightning speed, but our user protections do not.

The danger of trusting the pull of user demand to shape an industry is that users' short-term desires don't always account for long-term societal needs.

What Users Don't Demand: Regulation

The danger of trusting the pull of user demand to shape an industry is that users' short-term desires don't always account for long-term societal needs. Think of the personal choice of smoking versus its secondhand effects on other people, or the short-term savings of not carrying personal health insurance versus the long-term impact on public health, or the convenience of driving your own car to work versus the societal benefit of public transportation. In many situations, a user makes a choice and society bears the burden of it.

Now let's expand this dilemma to a uniquely modern one. Imagine you're a parent who wants to educate your child about technology, given the increasing need for young people to understand engineering concepts and have some familiarity with design. You purchase a cheap 3D printer and use it to impart lessons around technology, software, and manufacturing processes. You've brought into your home an amazing tool but also a potentially dangerous one.

For context, 3D printing (or additive manufacturing) is the process whereby a physical object is constructed using a 3D computer model and a standard machine that extrudes material to build the object, often layer by layer. These machines are extremely affordable for small-batch productions relative to the manufacturing equipment we've relied on until now. Most 3D printers can't yet create objects at the speed required for commercial scale, but flexibility was designed into their architectures from the beginning. Whereas the injection-mold manufacturing used in the last paradigm required specialized configuration, 3D printers are designed to enable someone to make almost any design a reality.

Today, 3D-printable items already range from the mundane, like plastic trinkets, to life-changing, like affordable housing. The first airplane with a 3D-printed part took flight in 2014. And the world's first 3D-printed heart was announced in April 2019. Simply put, 3D printing will democratize the production of *anything*.

On its face, this is amazing. Imagine completely eliminating the organ-transplant waiting list or not having to run to a hardware store when you need a nail. It's no wonder that hundreds of thousands of households have already invested in 3D printers. The world of home-printing critical goods is imminent.

Unfortunately, putting a modular manufacturing device in every household drives the same type of value-chain disruption that Facebook enabled with its publishing API. Customers are no longer beholden to the large companies that also were responsible for producing and distributing products. Instead, any amateur designer can use inexpensive computer design software to create models for production and then distribute their designs to millions of eager consumers by leveraging distribution networks of 3D-printer makers. With a simple download, end users can now fire them off to 3D printers.

Such modularization in manufacturing allows us to bypass the controls that have existed for generations in supply chains, regulated industries, and intellectual property. Relatively benign examples abound: Your child wants a new action figure—do you pay for it or just print an illegal replica? Much more serious, what if your driving-age teen puts a faulty home-printed part in your car? Even worse, consider firearms. Gun regulations vary across countries and US states, but they do exist—and many are enforced at the point of sale: What types of arms and ammunition can be sold and to whom? If anyone can download a model from the internet and print a weapon at home, much of our approach to gun control will be rendered moot.

Of course, most consumers bringing desktop 3D printers into their homes simply wish to take advantage of the flexibility of the new systems, not to forecast every potential use and failure of them. Users pull technology into their lives to scratch an itch: Facebook to entertain themselves and socialize, Lyft to get from point A to point B, 3D printers to educate their kids or get simple tasks done faster. Consumers don't (and shouldn't) be responsible for thinking about the implications of introducing new systems on the back of modular innovations.

As executives, if we rely on users to guide our ethical responsibilities, we are destined to be at best reactive—and, at worst, too late to chart the right course.

Luckily, if you believe that the internet will continue to enable rapid modularization in every industry, there are clear ways to navigate this compelling future.

Healthy Accountability

Around the time the news feeds debuted, Anne Wojcicki's 23andMe began offering direct-to-consumer DNA testing: Simply spit in a vial, and 23andMe would analyze more than 600,000 genetic markers to send you information about your health risks and ancestry. *Time* named it the Best Invention of 2008 for "pioneering retail genomics." And it was possible only because of the modularization in intellectual property related to genomics and gains in cloud computing that enabled high-volume storage, search, and processing. Of course, this modularization also created ethical gray areas.

Beyond empowering individuals with easy access to their health indicators, Wojcicki maintained a vision to accelerate and simplify medical research. The cost and time required to bring new treatments to market could be slashed with access to a sufficiently large, diverse database of consenting participants. It's easy to get caught up in the extraordinary possibilities. It's harder to consider tough questions about things like test validity, unexpected parentage discoveries, and the role of primary care providers in understanding results. It's tougher still to imagine all the new ways that access to this information might upend our existing social systems: What are your obligations to report a genetic marker for a disease to your health insurer? Can health insurers buy access to this information? What access should law enforcement have? What if you choose not to participate but your information can be easily inferred from that of a relative? And who's responsible for considering all of these questions and others?

Ownership and accountability are messy in the age of modularity.

Considering all possible societal implications is a big ask for people merely curious about their ancestry. And consumer genetic testing falls somewhere between the Centers for Medicare & Medicaid Services' regulations of clinical research (consumer DNA testing is not a clinical trial) and the Food and Drug Administration's regulations of drugs, biological products, and medical devices (the FDA now lumps consumer genetic tests in with medical devices).

Wojcicki spoke about this topic for four consecutive years at Stanford's Graduate School of Business. Her take is that, despite its challenges, trust is still crucial to keeping the health care system functioning. Therefore, if individuals couldn't contemplate the wide-ranging effects, and if regulators couldn't keep up with the breadth and pace of change, Wojcicki had to take responsibility to deliver that trust. Borrowing a proven concept from the existing health industry, she engaged an independent institutional review board to serve as ethical adviser on all of 23andMe's activities.

The fact is, 23andMe's data can be used for earth-changing research and, at the same time, have unexpected destructive effects. Skipping the middlemen of primary care providers in ordering genetic tests—or of clinical research organizations in collecting data—is not a question of morality but of how we as a society maximize the benefits while controlling costs. Pertinent applications of 23andMe's data will be debated, probably for years, before something like public consensus develops.

We've already seen that modularity enables businesses to quickly scale to entire populations, after discovering and delivering what users want—and that this speed shortcuts our long-standing approaches to public scrutiny. By seeking out third-party advisers to review the use of their data, Wojcicki has created a countervailing power to represent the societal viewpoint, just as any traditional research institution would maintain.

In redefining the way we access medical information and participate in research, direct-to-consumer genetic testing is another area where modular innovations could fail us without thoughtful action. The FDA, and certainly an individual consumer, cannot possibly consider all the positive and negative implications of merely spitting in a cup. The companies that find enormous value in this act must take on some of the ethical onus, as 23andMe has set out to do.

Intention to Action

Christensen et al.'s Theory of Interdependence and Modularity is a powerful explanation of how value chains evolve—and of the influence of consumer demand. As value chains split apart, innovators can reassemble them in response to customers' desires, in ways that take advantage of new technological options. Executives who embrace these changes should also seek to conscientiously address the often less-than-obvious ethical issues that arise. We suggest three courses of action:

1. **Assume you become the standard bearer.** Most innovators are comfortable playing on the margin. As disrupters who embrace modularity come up from below, it's easy for them to point to traditional businesses and refer to their ability to fulfill complex needs in the market. But success as a disrupter should come with a sense of obligation to change the paradigm, particularly when the upstart turns into the dominant platform. So instead of focusing only on the outcomes of your initial attack, work backward. Assume you become dominant. Then ask what is most likely to break, what can be done to prevent breaks, and how to handle them when they occur.
2. **Document the safeguards that would have prevented such failure in the current system.** Borrow a page from lean process improvement and start by mapping the complete value chain for the service you're providing as it existed before your company arrived. Next, chart out the future state in which you're dominant. Chances are, you've created an efficiency by removing or reducing the scope of some step. Learn the history of how this step evolved, and consider the safeguards ingrained within it: Are they regulatory? Are they related to standards? Are they social constructs? Consider the protections inherent in restricted access: What education or training did those with access have? If it helps, imagine how a horde of naive teenagers might misuse or misunderstand your service. Definitely contemplate how it may be used by malicious

actors. Safeguards have protected consumers as well as the market. Know them, and plan for how they will be addressed in the future state.

3. **Identify who is responsible for delivering these capabilities.** In some cases, it will be crystal clear: Ride-share services could not survive without trust in drivers, so Lyft and Uber must ensure background checks are done, even if they don't conduct them directly. In other cases, it won't be obvious: Are 3D printers "just a platform" facilitating exchange between model designers and consumers? Leaders need to anticipate that they'll be held accountable for the failures of the changes they usher in.

To put these recommendations into practice, it's important to assume success, understand the gaps, and take responsibility for the future that will be created. The particulars of implementation will vary by industry and company, of course. But we believe strongly that these three actions are key to recognizing where ethical uncertainties may arise from modularity and how to responsibly navigate that change.

The model for the Liberator, a 3D-printable plastic gun, was downloaded more than 100,000 times before a federal judge blocked the posting of 3D gun blueprints online.[3] Lucky for us all, not every household has a 3D printer; the printed parts have to be meticulously assembled; and, even when built correctly, the gun produced is unreliable (it's just as likely to misfire on its owner as on the intended target). In time, these complications will be worked out. But that also means there's time for regulators to plan for the obvious threat.

In other arenas, we should be more concerned. Industries such as lending, media, employment, and health care as we know them have evolved over the course of decades; their protections were sometimes hard-won and sometimes inherent in the very nature of the previous operators or target audiences. Faster than ever, disrupters and large corporations alike are reforming these value chains to take advantage of blazing-fast transfer of information, the application of artificial intelligence, and the creation of marketplaces and networks that distribute low-margin work. It's optimistic and reckless to assume that the existing protections will automatically port over to the newly modular systems.

Strict compliance with the laws, while crucial, is also insufficient to avoid the ethical pitfalls. In a piece for CNN Business, the former COO of Cambridge Analytica, Julian Wheatland, reflected on the scandal: "Cambridge Analytica made many mistakes on the path to notoriety. But its biggest mistake was believing that complying with government regulations was enough and thereby ignoring broader questions of data ethics and public perception."[4]

Lesson: The only rational solution is to embrace new ethical paradigms in a thoughtful way. Every executive should imagine the future that is bound to arrive and consider both the path toward consumer delight and the systemic protections that will be required.

5

Four Skills Tomorrow's Innovation Workforce Will Need

Tucker J. Marion, Sebastian K. Fixson, and
Greg Brown

Throughout history, new technologies have demanded step shifts in the skills that companies need. Like the First Industrial Revolution's steam-powered factories, the Second Industrial Revolution's mass-production tools and techniques, and the Third Industrial Revolution's internet-based technologies, the Fourth Industrial Revolution—currently being driven by the convergence of new digital, biological, and physical technologies—is changing the nature of work as we know it. Now the challenge is to hire and develop the next generation of workers who will use artificial intelligence, robotics, quantum computing, genetic engineering, 3D printing, virtual reality, and the like in their jobs.

The problem, strangely enough, appears to be two-sided. People at all levels complain bitterly about being either underqualified or overqualified for the jobs that companies advertise. In addition, local and regional imbalances among the kinds of people companies want and the skills available in labor pools are resulting in unfilled vacancies, slowing down the adoption of new technologies.

Before organizations can rethink how to design jobs, organize work, and compete for talent in a digital age, they must systematically identify the capabilities they need now, and over the next decade, to innovate and survive. For more than 10 years, we've been studying the impact of digital design and product development tools on organizations, their people, and their projects.[1] We've found that the competencies companies need most are business-oriented rather than technical. That's true even for brick-and-mortar companies that are trying to become more digital.

And most companies are beginning to realize that they can't just hire all-new workforces; there aren't enough qualified recruits, and the expense would be enormous. Instead, they need to retrain and redeploy existing employees and other members of their communities, in addition to hiring and contracting new ones to fill their needs. However, rapid technological change has rendered skill cycles shorter than ever; key competencies of even a decade ago are passé today, and most of tomorrow's jobs remain unknown.

Waiting for the fog to clear isn't an option. Companies must identify and develop the core skills their employees will need going forward. Our interviews, surveys, and case studies have revealed that most companies focus on refining the skills their people already possess, which doesn't prepare existing employees or new hires for the business challenges they'll face when using emerging technologies in their jobs. We've also found that young digerati, many of whom come into the workforce from narrow academic streams, are typically more captivated by digital technologies than they are by business problems. And yet, given the sweeping changes that the new technologies are likely to bring about, companies would do well to cultivate four broad business-oriented competencies in tomorrow's innovators.

1. Omniscience

To know it all may be a godlike, even insufferable, goal. But tomorrow's talent must aspire to understand everything—or at least much more than they currently do—about their businesses. Employees must grasp key connections: links between physical machines and digital systems, between each step of the value chain, between the company's current and future business models.[2] And they must know their customers' businesses—how and when their customers' products and services are used, how their customers' organizational processes work, and the related challenges and opportunities. That's the only way companies will be able to evolve from selling products and services to delivering outcomes—a process that will likely change the very businesses they're in.

For instance, a major medical device manufacturer we studied has moved from developing R&D-driven solutions to delivering patient outcomes, which has become possible because of new technologies and big data. The company needed to quickly employ more people with a systemic understanding of everything it does, including patient care and rehabilitation and treatment efficacy. To move the needle on patient outcomes, it's critical to understand all those aspects of the system and the associated variables. Thus, the business will demand that existing and new employees have a broader understanding about the underlying science, the delivery technologies, and the industry than almost all of them, other than top management, currently possess. Breadth of knowledge cannot substitute for depth, either; employees must also be able to make deep dives into the vertical aspects of the business when necessary.

Let's consider another example: The Canadian company Dental Wings is using recent advancements in digital design, digital imaging, and additive manufacturing, as well as a collaboration platform, to rethink its dental implant business. From the dentist's initial assessment to patient recovery, the company has started adopting new technologies to improve its processes and provide better care. For instance, all-new imaging capabilities provide more accurate pictures of the dental site that can be used not only to create digital models for implants but also to develop tools to help surgeons define the optimal surgical paths. That reduces exploration of the implant site, which helps reduce recovery time and lowers the risk of infection. To innovate at each step, Dental Wings' employees need to understand how the new processes and systems connect and work together.

The need to know more holds true for people in every function, but especially so in R&D and product design. In the not-too-distant future, product designers who are designing new earth-moving equipment will have to use AI and internet of things (IoT) sensor data to model, analyze, develop, and modify features in near real time. Once in the field, each prototype and its digital twin will operate simultaneously so that the designers will have access to data 24-7. They must be trained to use it to develop improvements for the current model on the fly as well as to better design the next generation of equipment.

In almost every brick-and-mortar company, dozens of digital platforms will have to be coordinated, the data mined, and the insights used in a harmonized effort between the human team and AI systems. Orchestrating all that data, whether from design outcomes or field performance, will require people who understand the value of each data point and how all the pieces fit together. It will also require knowledge across myriad disciplines, such as mechanical and electrical engineering, computer sciences, and product development, because the variables in a complex system interact in many ways. For instance, the location of a sensor on a suspension lever (a mechanical issue) will affect the data that the sensor electrically measures, which will in turn affect the mathematical algorithms that determine the lever's accuracy. Companies whose employees can manage and navigate complex data-based systems will be best equipped to improve the performance of their products, reduce maintenance costs, and attract and retain customers.

A Perfect Storm of Megatrends

Businesses tend to overlook the fact that the Fourth Industrial Revolution is gaining ground just as two other major shifts are exacerbating the skills shortage.

First, there's a demographic shift. With the baby boomer generation retiring and the working-age population declining in many countries, automation will likely replace many of the people who are leaving the workforce. Succeeding generations, such as the millennials and the centennials, seem to have different career aspirations than previous generations, as several surveys show.[3] Many would prefer to work for startups rather than incumbents. However, most large companies are old. Just 26 of the Fortune 500 companies were created in this century—like the centennials, who will soon constitute half the US workforce. These young workers have high expectations of employers, making it tough for traditional corporations to attract the young talent they need.

Second, as technologies change the way we work, they're creating a dynamic that differs from that of previous industrial revolutions. In the past, technology boosted the precision and productivity of workers with manual skills, enabling them to do tasks previously performed only by highly skilled and well-compensated artisans and craftspeople. Artificial intelligence and robots will have the opposite effect: They will increase highly skilled workers' precision and productivity but end up replacing many low-skilled workers, such as those on assembly lines, service desks, or maintenance teams. Even though some of those professions will survive, the necessary skills are changing fast: Miners will have to operate machines remotely, truck drivers will have to monitor self-driving rigs, and so on. Workers at all levels must learn to collaborate and coexist with learning machines.

---

2. Entrepreneurial Mindset

Although it may sound obvious, innovation teams will need to become more enterprising to succeed. They must become boundary pushers in terms of not just the products they wish to develop but also the processes they use. The two are closely linked.

In large businesses, R&D and product development teams are organized like most other functions. They must follow the company's guidelines about sourcing hardware, materials, and technologies to do their work and can use only IT-approved tools. R&D must adhere to time-tested procedures and rules for sharing information about or testing prototypes and product designs. And traditional R&D teams usually work in a centralized way, relatively insulated from the outside.

All that works well when business is as usual, but these are extraordinary times. R&D is meant to push technical boundaries, so R&D teams must learn to redraw organizational boundaries to keep pace with technological change. Essentially, they must become digital intrapreneurs, using the latest tools or, if necessary, creating them. That involves experimenting with new software and systems outside those recommended by IT, and even developing some solutions in-house.

For incumbents, that can be a shock to the system—most people are used to working on proprietary systems and tools, getting things "right" before launch, and offering better products over time. Moving toward open systems, beta versions, and constant iteration can feel like a clash of civilizations in established companies, but they need to do so to innovate for today, as well as tomorrow. Collaboration is central to this effort. One study of 152 managers found that companies that used digital tools for collaboration improved performance—as measured by the number of concepts and prototypes developed—during the early stages of innovation. And another study of 400 companies showed that more-innovative organizations, measured by similar yardsticks, used such tools more frequently than less-innovative ones. Since better collaboration leads to more innovation, the collaborative tools and processes that organizations use are critical. Figuring those out requires an entrepreneurial mindset as well.

For example, at a large company outside Boston, a new digital group is working on completely changing the way the organization designs products. This small team has asked for, and been given, the freedom to use any tools it wants, wherever they may originate. So the team has created a new system from scratch that allows it to test design structures in real time. The group also uses several digital platforms, most developed by unknown startups, to communicate and collaborate both internally and externally. It's unlikely that IT approves or is even aware of what's happening, but top management realizes that the company's digital transformation will never occur if teams like this one are confined by rigid boundaries.

There's a reason why entrepreneurs in high-tech startups are risk-tolerant, and it's time that intrapreneurs, or innovators in established companies, followed in their footsteps. Look at Proto

Labs, which manufactures injection molds and machined parts and offers additive manufacturing services. To accelerate the time it takes to develop the first tooling cuts for its clients, the R&D group quickly developed some software on its own. The program could identify possible manufacturing problems in the digital-parts files sent by clients.

Through its automated platform, Proto Labs R&D communicates any possible glitches it detects directly to clients so that they can rectify those well before production starts. If such revisions were made after test production had begun (as they were in the old days, before the homegrown software existed), the process would have been deemed client-unfriendly and would have cost both the client and the company time and money. Proto Labs has also added downloadable tools and other materials to help clients design better parts, ensuring that everyone in the ecosystem benefits from the process improvements. These offers are the outcome of entrepreneurial actions of Proto Labs employees.

3. Bottom-Line Focus

In a data-driven world, employees need to be just as skilled at thinking about business models as they are at designing and implementing systems. Thanks to IoT and other technologies, companies' value-capture strategies can be shaped not just by the marketing, sales, and business development functions, but also by R&D and product development. Design and innovation company IDEO's Tom Kelley describes people who look for business opportunities beyond the current challenges as cross-pollinators. Fostering that capability will be key.

Product engineers, for instance, must consider what kinds of sensors should be used, their placement, and the data types captured in light of possible revenue streams and cost savings. After all, big data poses as many challenges as opportunities. All hands must be on deck. The number of IoT-connected devices, estimated at around 2 billion in 2006, soared to 11 billion by 2019, and, according to Statista, is projected to touch 75 billion by 2025. Companies are capturing an enormous amount of data: IoT-generated data, estimated in 2016 at around 22 zettabytes (1 zettabyte equals 1 trillion gigabytes), reached 52 zettabytes by 2019 and is projected to hit 85 zettabytes by 2021.

While a company's digital people may appear to be on the front lines of the data explosion, they also need to be able to figure out what all that data means for the business and how it can be monetized. They must go beyond checking where the data originated, how dependable it is, where it is stored, and whether it has a coherent sequence. All that is useful but has become mere hygiene.

In focusing on business relevance, data technicians should be trained to ask some key questions: Can the data be used to monitor our products' performance *and* be offered as a service? Can that be done in real time? How else can the data be analyzed to generate insights about customers and their needs? For instance, can it be used to change the way customers schedule preventive maintenance for our products?

The need to be business-focused throughout the organization can lead to dramatically different customer-facing roles. One fast-growing company we studied develops sensor-based modules for the aerospace, automotive, and medical industries. It recently combined the roles of the product development manager and the product manager in all its lines of business—a radical step that immediately helped speed up cycle times.

To have a product position that is both inward- and customer-facing is unusual even today. Traditionally, the product manager would assess market trends and customer needs while developing working relationships with the company's clients. He or she would then feed the R&D team—led by a product development manager—the information to develop new products, systems, and solutions or improve old ones. Once the company combined the two roles, the speed with which new technical solutions were matched with prospects, and vice versa, rose dramatically.

Combing the two roles also created avenues for the cocreation of nontraditional solutions. For instance, by drawing on data from IoT sensors, the company was able to develop several new applications that reduced operating costs in areas that could not be assessed earlier, because the product development/product manager could now understand clients' pain points as well as all the solutions the company's technologies could provide.

4. Ethical Intelligence

Machines, overseen by smart humans, will make many design decisions. Though they are innately logical, they lack empathy. That will have consequences for companies, consumers, and society. Doing the right thing will become only more challenging as digital systems become increasingly complex.

People must examine machines' choices through an ethical lens—and weigh in. Companies will have to figure out how design decisions and digital systems affect each stakeholder and factor in the likely unintended consequences. In industries such as aerospace, automotive, and medical device development, traditional engineering processes like risk analysis and failure mode and effects analysis (FMEA) should also be deployed during the development of digital platforms and products. For instance, when Twitter's founders created the platform, they didn't imagine it could be used to influence elections with the use of fake accounts and bots. However, a coder putting the platform through a design FMEA would have identified the possibility well before people caught a glimpse of the platform's dark side.

Given AI's potential, every company needs to consciously decide what good judgment looks like. Take the case of Boeing's 737 Max 8, where, according to recent reports, pilots complained about an issue with the aircraft software while testing it, years before 346 people died in two crashes.[4] However, those concerns never made it to the Federal Aviation Administration—a tragic failure of ethics at all levels of the company. The countermeasures lie beyond the scope of this article but must include new codes of conduct, fresh corporate responsibility norms, KPIs that reinforce personal accountability, and specialized training.

To embed a watchdog mentality in the culture, companies should provide ethics training—and clearly define what *ethical* means in their specific context. Moreover, agility may be the norm, but companies still need to be disciplined in terms of process. That means a heightened emphasis on developing tools that improve quality and stop bad design from hurting people. Making processes more digital must not take away from the inherent value of techniques such as control plans and independent testing, whose importance should be engrained in tomorrow's talent.

As ecosystems develop, companies must use ethical intelligence to consider implications for all their stakeholders. At one open innovation platform, we found ethical breaches by the participants as well as the platform's management. The lapses affected the quality of ideas and input from the community as well as the trust among stakeholders. Companies must build guardrails into their platforms if they want to keep the faith of society, which already views corporations and intelligent machines with distrust. That could include more visibility into management processes and decisions, a clearer articulation of privacy policies, and better identification and reporting of anomalies in the system. Think of the impact on Facebook's image if it had reported the issues it experienced with foreign bots in 2016 in real time.

Why Structure Matters

Traditional companies will have to experiment with new organizational structures to get the best out of their people. Otherwise, tensions between well-entrenched managers and digital talent may thwart transformation, and the digital folks may walk out the door.

In their restructuring, it's important for companies to signal that digital transformation is critical for their futures. One radical approach is to replace the central R&D unit with a digital product design group. A well-known shoe company recently did this. The new group oversees the development of a new approach to product design, testing, and analysis, which will include customized generative design and analysis tools. Top management views this group as spearheading the company's future product development process.

Another option is to form a digital group that floats from project to project across the organization, as one leading consumer electronics company has done. There, digital experts hover over projects in various businesses and countries, providing input whenever asked or needed. The flexibility reduces the number of digital experts the company needs even as it helps retain them, because they enjoy the variety of opportunities and challenges the arrangement provides.

Some companies, like Apple, have internal venture teams to develop new products. Others are now doing so with a generational twist by creating new venture teams made up entirely of millennials and centennials to come up with new products and processes. A large pharmaceutical manufacturer we studied invited its youngest employees to conceptualize and implement a new way to connect patients, doctors, and the company during clinical trials for its products. Those employees used their native expertise in mobile technologies and social media to keep all stakeholders informed and involved. Top management let them run the show without allowing the rest of the organization to interfere. Funded by an internal venture capital panel, the project was tested, and eventually the company rolled it out to a wider audience. All too often, such projects are killed after their conceptualization, but companies that institutionalize entrepreneurial ecosystems can substantially improve their ability to innovate.

To be sure, the goal isn't to have a bifurcated talent pool in a company but rather an organization in which all the talent works together in a continuum, from hardware-focused experts to digital natives, from baby boomers to centennials. That's how many design and innovation companies now function, with older designers using sketches and hand-formed foam prototypes, while recent graduates go right to CAD software. Interestingly, the approaches can be effective if used together. At one design company we studied, the older designers, who preferred traditional methods, learned over time how the younger designers worked, and the younger ones gained a deeper sense of what they were doing from their older colleagues. It wasn't long before all the designers, regardless of age, were using digital tools for project management, communication, and collaboration.

It isn't easy for companies to change, especially from within. Kodak's middle management was skeptical of digital technology, for instance, and internal inertia was one of the key reasons it failed to make the transition from physical film.[5] However, identifying and bringing in the skills needed to move forward with innovation can help kick-start the transformation process. Indeed, doing so may make all the difference between success and failure.

Acknowledgment The research behind the ideas in this article was conducted with the support of PTC.

6

Education, Disrupted

Michael B. Horn

Employers are confronting sizable skills gaps in all parts of their operations, at all levels, and they can't seem to fill them by simply hiring new people. In today's tight labor market, there are about 7 million open jobs for which companies are struggling to find qualified candidates because applicants routinely lack the digital and soft skills required to succeed. In the face of rapid technological changes like automation and artificial intelligence, helping employees keep pace is challenging. And companies are wrestling with how to retain top talent—a critical differentiator in a hypercompetitive environment. No wonder a staggering 77% of chief executives report that a scarcity of people with key skills is the biggest threat to their businesses, according to PwC's 2017 CEO survey.

As a result, companies can no longer afford to wait for the traditional "system" to supply the workers they hope will help shape their future—the need is too acute and too urgent, particularly given that many higher-education institutions remain in denial. We must change how we educate both traditional college-age students and adult learners.

In 2019, when 4,500 people gathered at the ASU GSV Summit in San Diego to discuss innovation both in education and in talent development writ large, it was clear that the companies in attendance were eager to find alternative paths. At this conference, political leaders and policymakers join CEOs and venture capitalists to discuss the imperative of investing in human capital. Entrepreneurs in attendance work fervently to sell their wares or make deals, the likes of which have fueled a sharp increase in global mergers and acquisitions of education and talent development companies, up in total value from $4 billion in 2008 to $40 billion in 2018.[1] Executives from leading companies like Apple, Google, Facebook, Workday, and Salesforce.com attend to share ideas and learn about new ways forward.

The annual event provides a regular check-in on the state of corporate learning. In part, it's meant to coax companies to focus their efforts, because there's still a fundamental mismatch between how much they *say* they want to strategically invest in their current and future employees and what they actually *do*.

On a keynote panel last year, Leighanne Levensaler, the executive vice president of corporate strategy at Workday, bemoaned a great lack of investment in human capital despite all the buzz around the topic. Michelle Weise, the senior vice president of workforce strategies at Strada Education Network and the chief innovation officer for the Strada Institute for the Future of Work, has written that although 93% of CEOs surveyed by PwC recognized "the need to change their strategy for attracting and retaining talent," a stunning 61% revealed that they hadn't yet taken any steps to do so.[2] Employees seem to agree. According to a recent survey by Harvard Business Publishing Corporate Learning and Degreed, nearly half of employees are disappointed in their employers' learning and development programs.[3]

But there are some notable exceptions to this prevailing trend. For instance, in July 2019, Amazon announced that it would "spend $700 million over six years on postsecondary job training for 100,000 of its soon-to-be 300,000 workers." For now, Amazon says it intends to outsource that training to traditional colleges and universities. But once Amazon has begun to provide the bridge for that training, it's not hard to imagine that it will be well-positioned to create that training *itself*—without the "middle man" of colleges and universities—in the future.[4] Although Amazon's competitors will undoubtedly keep a close eye on its training moves, perhaps the education industry ought to keep an even closer eye, given that those moves may herald a total transformation in the landscape of learning, from college through retirement.

To put this development into perspective, it's worth stepping back to consider how learning has already evolved in recent years, before situating Amazon's announcement within the broader opportunities and challenges facing employers.

What's Next for Adult Learning?

Even before the coronavirus pandemic struck the United States in 2020, education was in the midst of digital transformation.

That this is true is no longer hotly debated. Online learning emerged over two decades ago as a technology category that enables a range of potentially disruptive business models. No longer do students need to convene at a central location to enjoy a real-time, interactive experience with a teacher and peers. They can instead participate from anywhere in the world, in a more affordable and convenient fashion.

This trend is growing rapidly in postsecondary education. Today, roughly a third of students in the United States take at least one online course as part of their accredited higher-ed experience, and over 15% study exclusively online.[5] Many of these students are adults who are employed while they learn. Countless more workers take supplemental courses on platforms like Coursera, Udemy, and edX.

Indeed, online learning has led to the creation of numerous organizations and offerings that support companies' talent development efforts. For example, Pluralsight, LinkedIn Learning (built on the acquisition of Lynda.com), Learn@Forbes, and Udacity help employers re-skill the workforce in myriad areas, often in specialized or cutting-edge fields. Startups like Guild Education and InStride allow companies to work with colleges and universities to offer learning as a benefit. Degreed has emerged to measure and help assess the learning and skills inside an organization. Coding and engineering boot camps like General Assembly and Galvanize and other so-called last-mile education providers (many of which offer blended or fully online programs) are increasingly working directly with enterprises. And universities like Arizona State, Bellevue, Southern New Hampshire, and Ashford, as well as schools like Ultimate Medical Academy, are partnering directly with companies such as Starbucks and Walmart to offer education to employees.

The pace of innovation in corporate learning is frenetic—and highly uneven. As providers compete to serve enterprises, there is not one monolithic answer for what corporate learning will look like in the future. Just as companies have always patched together a variety of learning solutions to support their needs, they will most likely continue to do so.

But what this abundance of new approaches and players has led to is the same thing that disruptive innovation has wrought in countless other fields: far more affordable and convenient options. In the case of learning and talent development, such offerings have the potential to allow companies to make more significant investments in their greatest asset: their employees. Which companies will leverage this opportunity to improve both their bottom lines and the welfare of their people? The answer to that is not yet clear, although it will be interesting to see whether a critical mass of organizations will follow Amazon's lead.

An Interdependent Solution to Training

In many ways, Amazon's announcement shouldn't have been a surprise. The need for better-trained talent is clear in companies across the globe, and Amazon is taking a somewhat predictable path.

Amazon's efforts resemble what we've seen happening in other technology arenas for decades, bearing out Clayton M. Christensen's Theory of Interdependence and Modularity. The theory tells us that in the early years of a new paradigm, in order to succeed, product and service providers must integrate across all the unpredictable and performance-defining elements of the value chain. Think of how, in the early days of mainframe computers, IBM integrated hardware manufacturing with the design of interdependent operating systems, core memory systems, application software, and so on. IBM recognized that to thrive, it had to do much more than make machines that would play nicely with modular components created by others. It had to own the whole value chain.

We are now entering a similar moment in workforce education. The status quo that existed in the industrial economy and the early years of the knowledge economy—in which the links

No longer do students need to convene at a central location to enjoy a real-time, interactive experience with a teacher and peers. They can participate from anywhere in the world.

between companies and the educational institutions that fed them were predictable and good enough—is no longer adequate.

In the case of Amazon, the step in the value chain that's not good enough is the education that colleges and universities provide. Because the subject matter Amazon's employees need to know is changing rapidly and building the curricula through traditional higher-ed faculty and processes would be too cumbersome, Amazon has concluded that it will in essence take a much more active role in the education and training of 100,000 of its employees. What may be equally interesting to monitor is where Amazon goes with this development. The company was its own first customer for Amazon Web Services before opening up that offering to others. It's not hard to imagine Amazon doing something similar for corporate learning. Will Amazon shape the future of the global workforce through its own education programs? The company's timing, it would seem, couldn't be better.

A Focus on ROI

For corporations to invest in learning solutions in a sustainable way, there will most likely need to be a clear and compelling return on investment. As Allison Salisbury, a partner and head of innovation at education venture studio Entangled Group, has observed, companies can take at least five different angles when investing in human capital: providing on-ramp programs, upskilling, re-skilling, outskilling, and education as a benefit.[6] Some of these approaches may be more sustainable than others, but each one has a distinct ROI. For instance, on-ramp programs bolster the quality and diversity of candidates for hard-to-fill roles by offering short-term training that creates a direct pipeline for employers. Outskilling programs, which are growing, help employees who don't have a future at a company build a skill set to change careers. Companies offering such services become more desirable places to work and enhance their reputations in the labor market.

In today's economy, the imperative to invest in many, if not all, of these categories is evident for employers. Companies are competing for a scarce resource: people qualified to execute mission-critical tasks. Hence the Amazons and AT&Ts of the world are announcing major half-billion-dollar-plus bets on training.[7]

But are these just fair-weather investments? Given the state of the economy in the wake of the pandemic, which of these categories will companies abandon? If the past is any guide, the most vulnerable categories will be those where the returns are the least direct—areas such as outskilling, perhaps, where the immediate benefits to the company are more reputational than financial. Even upskilling will probably be at risk—despite its obvious economic upside, given the widely acknowledged skills gaps that businesses urgently need to fill—unless employers can show a clear ROI that is *better* than other potential investments in automation and the like, as Mike Echols, formerly the director of Bellevue University's Human Capital Lab, has written.[8]

The Measurement Challenge

The biggest challenge for companies that want to invest sustainably and heavily in human capital may lie in figuring out what kinds of people they need. For all their apparent sophistication in data analytics, few employers have a clear sense of the underlying skills, competencies, and habits of their most successful employees—never mind their future workforces. As a result, they

Companies are competing for a scarce resource: people qualified to execute mission-critical tasks.

don't know what to look for when they post jobs, interview candidates, and hire new employees.

A key sign of the imprecision of the hiring process is that nearly 50% of newly hired employees fail within 18 months.[9] And that failure has significant costs—$15,000 on average, according to a CareerBuilder Survey.[10]

Why do employers struggle to understand what is important to succeed in certain positions? Partly, it's because experts are notoriously bad at knowing what they know. According to the book *How Learning Works*,[11] as individuals gain expertise in a particular role or field, they go through stages, from novices who *don't* know what they don't know to novices who *do* know what they don't know to experts who *know* what they know to experts who *don't know* what they know. The reason is that automating knowledge—essentially moving it into an individual's subconscious as background information—is critical to freeing up space for the complex and creative tasks that an expert performs. As a result, though, asking top performers in a company to write a job description, for example, or to say precisely what skills are at the heart of correctly doing a job, is not as simple as it sounds, because the experts literally don't recall. They are good at their jobs *because* much of their knowledge has been automated, so they aren't able to easily articulate what skills are essential. What's the solution to this problem?

For years, one of the most trusted ways to identify key competencies was cognitive task analysis, a process of observing and documenting the underlying activities involved in performing a job. But cognitive task analysis is relatively costly and time-consuming, so most employers don't do it.

Herein lies an opportunity—and so a wave of providers is sweeping in to offer new ways to measure the skills of employees.

Degreed, for example, has built a platform that records all the learning employees do, in an effort to understand their various learning pathways. It also offers a range of skill assessments to certify experts in various domains. LinkedIn Learning offers similar assessments, along with learning software to help people upskill, and tracks people's self-reported skills and their connections to various jobs.

If players like these are successful in capturing the real skills at the heart of work and measuring their attainment, that could translate into more precise measurement of the return on investment in human capital. And that could, in turn, lead employers to take far better advantage of the emerging slate of disruptive tools dedicated to helping people learn in a sustainable and strategic way rather than an episodic and ad hoc way.

7

Betting Big on Employee Development

Ardine Williams, interviewed by *MIT SMR*

Talk about how technology will affect the workforce of the future, and Amazon is likely to enter the conversation. In July 2019, the giant retailer announced that it would upskill 100,000 employees—a third of its US workforce—over the next six years by spending as much as $700 million. Leading the initiative is Amazon HQ2's vice president for workforce development, Ardine Williams, who has 35 years of product development, marketing, corporate business development, mergers and acquisitions, and HR experience in the high-tech industry. In an exclusive interview with *MIT Sloan Management Review*, Williams, who began her career as a US Army officer, explains the rationale for the upskilling initiative and the benefits to the business, local communities, and individual workers.

***MIT Sloan Management Review*: Why is Amazon doubling down on workforce development now?**

Ardine Williams: Amazon believes that it has an important role to play in the creation of good jobs. Good jobs have three elements: (1) a good wage—that's why we announced a $15-per-hour minimum wage across the US in November 2019; (2) robust benefits from the day you join—Amazon's fulfillment center associates have the same benefits today that our executives do; and (3) the opportunity for people to create a career by gaining experience and building skills that give them more options to progress over time.

We believe that technological advances will continue to change job content, so upskilling will always remain an important component of Amazon's workforce development.

How do you perceive technology change and how it's affecting Amazon's workforce needs?

As long as human beings have worked, technological advancements—from wheels to steam-powered looms to computers—have changed how they work. Having said that, the challenge today is different in two ways: The pace of technological change has accelerated, and the impact of technology on jobs is being felt at the task *and* skill levels.

Jobs typically don't disappear because of technological change, but [such change] often creates new opportunities or changes the nature of what people do. For example, in warehouses where Amazon has deployed automation technology, workers have gained opportunities to acquire new skills. Through retraining, employees can learn to clear the lanes where robots operate, reset robot paths, and perform basic maintenance and repairs on robots. Upskilling ensures that employees can continue to contribute to the company.

It's important to be deliberate in planning what work will fill employees' newly available time, the skills they will need to complete that work, and what the company will do to equip them with those skills.

Since Amazon attracts the best talent from both the physical and the virtual worlds, does it really need to invest so heavily in retraining and upskilling? To what extent is this initiative all about brand-building and demonstrating social responsibility?

Branding is important to all employers, make no mistake. That's how companies attract and retain smart and passionate people, who must work together to innovate for customers. But Amazon's upskilling initiative is about helping people grow careers through work experience and then building on that by adding skills through training.

A combination of work experience and new skills creates career momentum. That momentum may be vertical, up a traditional career path in functions such as finance or accounting, or it might look more like a lattice that takes you from one function to another, such as from recruitment coordinator to project coordinator to project manager. We'd love every employee to build a career with us, but in many cases, career progression will take people away from Amazon. That doesn't mean we can't, or shouldn't, play a role in their development.

What kinds of skills are most attractive to participants in Career Choice, Amazon's upskilling program? Are they all tech-related? How exactly does Amazon identify the shortages in local labor markets?

Career Choice provides training for in-demand jobs that pay more than Amazon does in local communities, and it offers employees new career paths. The program trains for about 37 distinct job types in five general families. Our three most popular career fields, in random order, are medical, IT, and transportation (more specifically, commercial drivers). It's tough to argue that those three fields aren't technical.

There are four key steps I see in developing Career Choice programs: identifying roles that are in demand locally; finding employers that are seeking skilled talent; understanding specific skills requirements; and working with training providers, such as community colleges, to tailor programs for full-time workers.

We use the US Department of Labor statistics and third-party data to identify jobs that are trending. But the data is always lagging, and there isn't a single source of truth for labor shortages or for the knowledge, skills, and capabilities that companies need. Those factors make it challenging to scale our efforts.

Employers are usually reluctant to invest in training that could end up benefiting rivals when people leave. Amazon doesn't seem very concerned about that. Why?

Whether Amazonians choose to build their careers with us or go elsewhere, we want to help them succeed. That isn't philanthropy; it makes good business sense.

Even when people leave us for other employers, they fill jobs in local businesses that would otherwise go unfilled. Production increases, the pipeline of qualified talent helps local businesses succeed, and it attracts new businesses to the area. Discretionary incomes go up. We're a retailer, so we want those people to continue to shop at Amazon and to think of us as a good partner in the community.

Are Amazon's workforce development programs all created and delivered in-house, or do you use outside vendors?

For Career Choice, the upskilling initiative, our providers are predominately community colleges and other specialized third parties that deliver career-specific training. Other programs, such as Amazon's Machine Learning University, are created and taught internally, and our US Department of Labor–registered apprenticeships are a combination of internally and externally developed programs.

Why does Amazon think it's necessary to conduct or manage the retraining itself? Do traditional forms of education fall short of developing skills, focus on the wrong skills, or adapt too slowly to develop them?

We try to work back from the customer: in this scenario, the employees who are upskilling. Many of our programs are built by Amazonians for Amazonians, with some leveraging digital training, courses, and certifications with partners. Sometimes it makes sense for an Amazonian to lead the teaching—at Amazon Technical Academy, for example—so that what is taught can be immediately applied by employees in their roles in the company. We have hundreds of employees designing and building training programs, and we still have hundreds of job openings in training and development.

What kinds of partners do you typically work with, and why? Is it tough to collaborate with community colleges?

It all depends on the program. We work with community colleges and experts, but to be effective, we need to meet employees where they are. We want to make it easier for them to gain the skills that they need to grow their careers. So wherever possible, we bring the training into the workplace.

Community colleges have excellent programs, but those are usually full time. Adult learners in jobs with good benefits usually face the dilemma of whether to quit their jobs to take the training to get a better job—or to stay put. In many cases, people don't have the luxury to stop working to go to school. That's why we've partnered with community colleges to make part-time programs, bring them onsite into our fulfillment centers, and offer them at shift-friendly times during the working day. As our employees complete those programs, we schedule job fairs so that they can find new jobs. We expect to have 60 Career Choice classrooms by the end of 2020. The classrooms are important; they make upskilling both convenient and visible.

What do the credentials of the future look like?

There's no doubt that the internet has made education more easily accessible. A smartphone and a broadband connection are all that's required to make content available where and when it's needed. Our next challenge is figuring out how to identify the best content and certify competencies in critical skills. There's a lot of innovation happening in that space.

Stackable industry credentials are gaining in popularity. In some cases, requirements are clearly defined. Think CompTIA and Cisco networking credentials, for instance, and state licenses for medical professionals. In other cases—for the data analyst, cybersecurity expert, or app developer, for example—a wide range of credentials and certifications is available. So it's difficult for students and employers to know which courses and certificates lead to job-ready skills.

Preparing graduates of two- and four-year colleges and universities to be job-ready is a shared responsibility. Employers and educators must work together to ensure that the knowledge, skills, and capabilities that work demands are clear.

The Greater Washington Partnership's Capital Collaborative of Leaders in Academia and Business (CoLAB) is a great example. Through that initiative, business and academic institutions are working together to develop the workforce that the region needs today and tomorrow. CoLAB recently launched a data analytics certification program that identifies students who have completed an industry-approved curriculum in this high-demand area. Employers and educators agreed on the required knowledge, skills, and abilities. Then, educators chose the best pathways for their institutions; for example, they could offer data analytics as a minor, an area of concentration, or a series of electives. The first cohort will graduate next year, and we are all working together to build the metrics to measure success and provide feedback.

Did you conduct some pilots before scaling Career Choice? How did you measure their success?

The Career Choice program is actually around 8 years old. We've learned, and we continue to iterate and invent. In the early days, the program was more akin to a traditional tuition assistance program, and eligible employees could select any course of study. We had a lot of uptake but struggled to demonstrate that the program was really preparing our employees for new career paths.

At that point, we decided to work back from the end goal: We began identifying the roles that would provide the greatest likelihood that upskilling would lead to new careers. While that helped us start moving in the right direction, we were still too far downstream. Community colleges' success metrics are enrollment and completion, and our program goals were post-completion employment and incomes. We needed to focus on identifying in-demand jobs in the local community that paid more than we do.

That's when we realized that we needed to work across the entire ecosystem—educators, employers, and catalytic agencies—to link education to employment. We also work with third-party conveners, such as industry associations, that have a broad view of talent gaps in local industries. We use a variety of metrics to track our progress but still have some work to do on that front.

How does Amazon's organizational culture support workforce development?

We don't believe an employee needs to follow a specific career path in the company to be successful. Ours is a culture of builders; we look for candidates who are curious about cause and effect, and who are passionate about rolling up their sleeves and working together to innovate. Our leadership principles—such as Think Big, and Learn and Be Curious—can be applied to employees exploring training programs, too. From machine learning to medicine and transportation, if someone is interested, there will always be an opportunity at Amazon to explore it. Amazon doesn't require employees to stay in a role for a minimum period before transferring to a new position. Nor is there a requirement that people must remain employed with Amazon after completing an upskilling program.

What do you think Amazon's workforce is going to look like in, say, 2030? What new skills will employees need?

The line between tech and nontech, or between STEM and non-STEM, is blurring rapidly. Earlier, technology was isolated; now, it's almost impossible to find a job that doesn't have technology infused in it—and that's the real disrupter. We can no longer say "STEM or"; it is "STEM and." Successful employees will have to be technically literate. The more companies can do to educate the workforce about technology, the more prepared they will be to deal with the changes in jobs that are coming down the road.

Even careers that have been traditionally identified as nontechnical will require foundational technical expertise tomorrow. We will need people who can identify the problem to be solved from a sea of data, communicate that problem cogently, assemble the right team to tackle the problem, work collaboratively with a team that has a wide range of skills and abilities, and bring the right technology to bear in order to innovate for customers.

Does Amazon plan to roll out its workforce education programs to other companies as a business venture?

The American workforce is changing. There's a greater need for technical skills in the workplace than ever before, and a huge opportunity for people with the right skills to move into better-paying jobs. The purpose of Amazon's upskilling initiative is to ensure that our employees can migrate into more-technical roles, at Amazon or elsewhere, through training and career support. We share our Career Choice program for free with other companies and learn from them so that we can increase the impact of our programs.

III

The Competitive Element

8

From Disruption to Collision: The New Competitive Dynamics in the Age of AI

Marco Iansiti and Karim R. Lakhani

Airbnb is colliding with traditional hotel companies like Marriott International and Hilton. In just over a decade, the online lodging marketplace has assembled an inventory of more than 7 million rooms—six times as much lodging capacity as Marriott managed to accumulate over 60-plus years. In terms of US consumer spending, Airbnb overtook Hilton in 2018 and is on track to move ahead of Marriott.[1]

Although Airbnb serves similar consumer needs, it is a completely different kind of company. Marriott and Hilton own and manage properties, with tens of thousands of employees in separate organizations devoted to enabling and delivering customer experiences. And whereas the two traditional lodging companies are made up of clusters of different groups and brands, with siloed business units and functions equipped with their own information technology, data, and organizational structures, Airbnb takes a radically different approach: Its core function is to match users to hosts who have unique homes or rooms to rent on a daily basis, via its platform. In the process, Airbnb accumulates customer data, mining it for insights and to produce predictive models to inform key decisions. It is often able to give its customers a superior experience, with far fewer employees, than its hotel-industry competitors.

Airbnb is representative of a wave of new organizations that are built on an integrated digital foundation. Every time we search Google, buy from Alibaba or Amazon, or get a ride from Lyft, the same phenomenon occurs. Rather than relying on traditional business processes operated by workers, managers, process engineers, supervisors, and customer service representatives, these companies deliver value through software and algorithms. Although humans design the systems, the computers do the work: producing search results, setting prices, identifying and recommending products, or selecting a driver. This reality defines a new kind of digital company, with data and AI at the core and human labor pushed to the edge.

The Analysis

- The authors conducted several research projects to understand and model the impact of network effects, digital platforms, and digital learning on company performance and competition.
- They have also led research projects across more than 500 organizations to understand the impact of analytics, digital operating models, digital networks, and AI.
- They have advised many organizations on these topics, including Amazon, Disney, Facebook, Fidelity, Marriott, Microsoft, and Mozilla.

Many of these changes are being played out in other parts of the economy as well, including the retail and entertainment media sectors. The collisions between innovators and established players are forcing leaders of existing companies to reexamine how they do business in environments where new players follow radically different rules. In many settings, making small or incremental changes won't be enough. Rather, companies will need to fundamentally alter how they gather and respond to information and how they interact with their customers and users. Organizations will have to rethink their operating models from top to bottom.

The digital model has intrinsic advantages over traditional models. Thanks to its operating architecture, Airbnb, for example, can take advantage of network effects in its platform, and learning effects through its data integration and AI systems, to rapidly improve operational scale, scope, and learning. Whereas Marriott's ability to grow and respond is limited by traditional operational constraints, Airbnb digitizes internal processes and connects beyond the company boundaries to build an ecosystem of travel services. On an ongoing basis, it can mine its data to acquire new customers, identify traveler needs, optimize experiences, run experiments, and analyze risk exposure. Along the way, it can accumulate even more data on hosts and travelers and use artificial intelligence and machine learning to gain new insights. Beyond the lodging business, Airbnb is expanding the scope of its offerings to include other types of travel experiences, such as concerts, cooking classes, and local tours, opening its ecosystem to a variety of new service providers.

Airbnb isn't the only company leveraging its digital capabilities to drive change in the global travel market. Other well-known travel brands like Booking.com, Kayak, and Priceline (all owned by Booking Holdings) also use software- and data-centric operating models to promote scale, scope, and learning without encountering traditional operational constraints. In November 2019, the public valuation of Booking Holdings was almost double that of Marriott.

The entire industry is transforming before our eyes. In just a few years, both Airbnb and Booking have dramatically increased the number of room nights sold and have catapulted into leadership positions. Market concentration among the leading traditional hotel operators is also increasing, with merger-and-acquisition activity on a high boil. Marriott, for example, merged with Starwood in 2016 to exploit synergies across their loyalty programs and related data assets. In a race against time, Marriott is working hard to re-architect its operating model to remain competitive against Airbnb's and Booking's data-driven growth machines. Indeed, the entire lodging and travel industry is in the midst of major upheaval, with companies like Marriott and Hilton in a fight for their existence.

The Competitive Dynamics of Collision

The collision between digital and traditional companies shows what happens when user needs are met by a new kind of operating model that digitizes some of the most critical tasks to deliver value. In the travel industry, customer needs haven't changed—travelers continue to need accommodations and experiences. But unlike hotel chains, Airbnb's and Booking's systems can satisfy those needs without armies of hotel managers and salespeople or cumbersome labor- or management-intensive operating processes.

In many ways, Airbnb and Booking are built like software companies. They provide a software layer to the travel industry, functioning in effect as operating systems. If Marriott is the industry's IBM mainframe company, Airbnb and Booking are vying to become the Windows operating system. In doing so, they aim to push traditional operational bottlenecks outside the walls of their organizations and remove constraints on their own scalability, scope, and learning potential. This dramatically shapes their ability to deliver value to customers. Traditional businesses can scale up quickly but often run into diminishing returns in their value generation as they encounter problems from getting too big. They face diseconomies of scale in human-centric managerial processes and administrative inertia, which slows their growth and, if they are not careful, can lead to worse outcomes.[2]

Digital operating models scale differently. Google's search engine and Alibaba's Alipay payment app, for example, can scale to a virtually infinite number of customers, link to a vast array of complementary businesses, and get better with experience and with more users, because they do not suffer from any diseconomies of scale. Companies with traditional operating models encounter diminishing returns as they scale and grow the number of customers they serve, but those with digital operating models can achieve increasing returns to scale. The collision occurs when the value curves of traditional and digital operating models intersect. (See "A Collision in Action.") Although nothing grows forever, and the value generated by digital operating models will eventually plateau during the period when managers and executives of traditional incumbent companies need to react, the scale potential can seem unlimited. Indeed, the growth of some of these digital operating models will slow only through a catastrophic failure such as a massive privacy scandal or a cybersecurity breach, or through regulatory concerns about market concentration and consumer data protection.

The travel industry examples show how AI, learning, and network effects can go hand in hand to build a rapidly growing value proposition in a series of self-reinforcing loops. As the operating model develops more connections, it also develops new opportunities to generate and accumulate data. With more

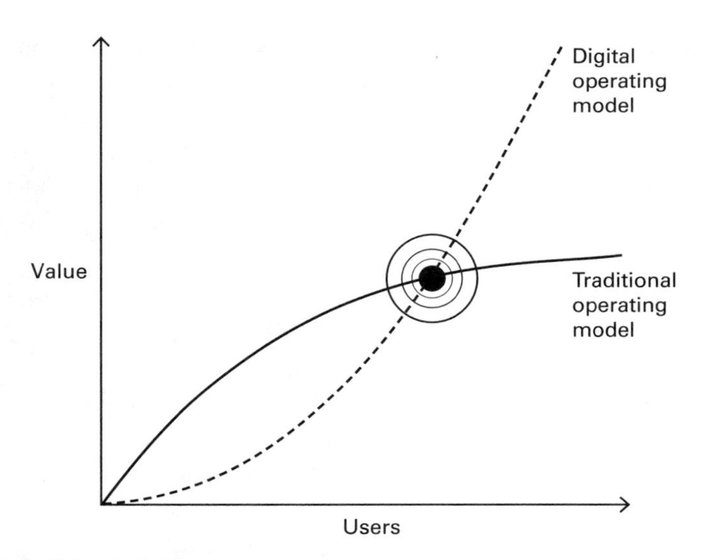

A Collision in Action

Traditional and digital operating models collide with one another where their value curves intersect. While the former tend to have diminishing returns, the latter can continue to grow in scale, scope, and learning, increasing in value as users and engagement grow.

data come more opportunities for better services and greater incentives for third parties to plug in. This, in turn, increases the potential for learning and amplifies network effects. In general, the larger the network, the more data it generates, the better the algorithms, and the higher the value it can deliver.[3]

These self-reinforcing loops in network and learning effects make a big difference to the nature of competition. In traditional operating models, the value that can be delivered begins to level out as the organization grows. This often implies that entrants can threaten incumbents, because the advantages of scale are significant but not insurmountable. New companies can bring innovative solutions to market even on a smaller scale—think of a network of boutique country inns taking room nights away from Marriott resorts. In contrast, in digital operating models, traditional constraints go away, and the value delivered will continue to increase, possibly at a faster and faster rate. No small-scale outfit can reasonably compete with Airbnb.

This has an exponential competitive effect. As digital operating models deliver more value, the value-capture space left for traditional players shrinks, making it increasingly difficult for traditional companies to sustain a profitable offering. Airbnb and Booking do not compete head-to-head with Marriott or Hilton by opening their own hotel chains. Rather, they extract much of the consumer value and commoditize the hard-won brands and experiences of the hotel companies. While hotel companies may never disappear, their profits will continue to migrate to the "software layer." For example, research shows that Airbnb interferes with the ability of hotel chains to protect their prices during busy time periods (for example, when a special event like a convention or the Super Bowl comes to town); by increasing the supply of alternative beds, Airbnb puts a ceiling on the prices that hotels can charge, to the benefit of consumers and the detriment of hotels' bottom lines.[4]

A Repeating Pattern

The Airbnb story is becoming a common one—many of its themes are being played out in other industries. Just as the cloud computing services of Amazon and Microsoft are replacing traditional IT software and hardware solutions, and fintech providers such as Wealthfront and Kabbage are nipping at the heels of established banks and investment firms, marketplace platforms such as Alibaba, JD.com, and Amazon are overtaking traditional retailers. The transformations are profound, with serious implications for how companies design their business models (that is, how they create and capture value), how they execute their operating models (how they deliver value), and the competitive dynamics and market structures of their industries.

Below, we will discuss in more detail what's happening in the retail and entertainment media industries.

Retail

Amazon was founded in 1994 and was among the first online retailers, establishing a pattern for other online retailers, including Drugstore.com, JD.com, and Pets.com. Over time, the online retailers created platforms, and Amazon broadened and deepened its marketplace with thousands of third-party merchants offering millions of products. In essence, Amazon became a scaled-up Sears and Kmart, but without needing physical stores or having to carry extensive amounts of inventory.

Traditional retailers were able to compete with the first generation of online retailers fairly well; the big changes didn't occur instantly. For example, the ability of online retailers to tap into data and analytics was still quite limited, and like others they had to suffer through supply chain bottlenecks. Some online retailers (Pets.com and Drugstore.com, to name two) proved incapable of meeting customer needs any better than traditional retailers and went out of business.

However, Amazon found a way to take on traditional retailers using a data-centric operating platform to transform the retail experience. The transformation went beyond simply moving transactions online. It called for a fundamentally different

Traditional retailers were able to compete with the first generation of online retailers fairly well; the big changes didn’t occur instantly.

operating approach, based on a data- and AI-centric analysis of the customer in order to personalize the retail experience. Retail supply chains became centered on software, shifting labor from the core of the process to the edge (for example, in picking products from warehouse shelves), which removed traditional bottlenecks and scale constraints. By the late 2010s, the weaknesses of traditional retailers were in full view, illustrated by the demise of many well-known players, including Toys R Us, Sports Authority, Sears, Nine West, Kmart, and Brookstone.

It took a while for online retailers (notably Amazon in the United States and Alibaba and JD.com in China) to figure this out and deploy the right operating model, but once they did, traditional retailers faced challenges like never before.[5]

Entertainment

The earliest data- and software-centric operating model to collide with traditional players in the entertainment industry was Napster in the late 1990s, which allowed people to digitize and share their music online—skipping over the usual payments to the various players in the music industry and offering music as a "free" service. Despite its immense popularity, Napster ran into a buzz saw of legal troubles that led to its shutdown in 2001. Following Napster's demise, Apple Music, Spotify, and others clashed with traditional music-distribution companies, eventually transforming both business and operating models for music distribution in the United States and beyond. Essentially, they converted a music-acquisition expense that individual consumers made on a case-by-case basis (resulting in a limited home-based music library) into monthly subscription services, offering unlimited music anywhere, anytime. Spotify, YouTube, and Apple are now the main hubs for music flow in the United States and Europe.

A similar battle has taken place in video. Although RealNetworks launched the first internet streaming video company in 1997,[6] it soon attracted stronger competitors such as Microsoft and Apple, and eventually YouTube and Netflix. YouTube and Netflix offered more compelling value propositions for consumers, as well as more scalable operating models based on software, data, and AI. However, the video market shows that despite similarities in the operating models, significant differences in business models can lead to differences in competitive outcome.

YouTube, with a business model based on aggregating a huge community of small content providers, dominates video sharing. By taking advantage of strong network effects, it has become a true video-sharing hub. In contrast, the kinds of premium video-streaming services Netflix provides originate from a more concentrated set of professional content production studios. Although Netflix's data and learning advantages are important, it can't compete with YouTube's network-effect advantages at scale, which are gained by the video-sharing company's ability to aggregate content from a vast variety of sources. This weakness has permitted a number of companies, notably Hulu, Amazon, and Apple, to also focus on content production and compete directly with Netflix. Without access to strong network effects, these providers are attempting to differentiate themselves by tapping into a more focused range of unique content (through special studio relationships and vertical integration).

As a group, Netflix, Apple, and Amazon are also colliding with traditional cable and satellite television providers, as well as traditional TV and entertainment companies, providing over-the-top (internet-based) video content distribution platforms that have rapidly scaled to hundreds of millions of users globally. Threatened by more efficient data- and AI-centric competitors, and mindful of the devastation that has occurred in other industries, traditional media companies are scrambling to react, merging with content and internet service providers to spark transformation, and re-architecting their operations around a digital core. Digital cable provider Comcast has made major headway by introducing and upgrading its Xfinity X1 platform. Disney is following suit with its ESPN+ and Disney+ streaming services. In contrast to video sharing, the premium content-streaming setting is likely to be highly competitive for the foreseeable future.

The changing shape of the entertainment industry highlights some interesting issues. As we have seen in other contexts, being first offers no guarantee of success. And the transition to a digital operating model is pervasive throughout the entire industry. Both new and old competitors must shift to an operating architecture focused more on data, AI, and digital networks. Finally, despite convergence in the operating models, different players can still achieve different kinds of competitive outcomes (as we have seen with video sharing versus the creation of premium content) because of the nature of each business model and the strength of network effects available.

How Collision Differs from Disruption

Collision and disruption are, of course, closely related. They are connected historically through a "law" named for computer scientist Melvin Conway, who noted that organizations are constrained to perform activities (design, in the original example) that reflect the communication patterns prevalent in each organization.[7] Conway's law explains why the physical architecture of products or services developed by companies reflects their organizational architectures. If we look at the organization of a product development project, we will see separate groups dedicated to the design of each component or subsystem. But because this architecture makes it easier for organizations to perform similar tasks over and over again, it also makes it difficult for them to respond to change, causing organizational inertia.

In a landmark 1990 paper, economists Rebecca Henderson and Kim Clark argued that "architectural" innovations—ones that require changing the architecture between technological components—are a particular danger to established companies.[8] The paper explained the demise and subsequent obsolescence of many notable companies that failed to change their organizational architectures to match the new requirements. Among them: RCA's failure to re-architect and miniaturize its tabletop radios and music devices even in the face of competition from Sony (which licensed RCA's technology!), and IBM's failure to transition from mainframe computers to PCs.

The idea of architectural inertia, in turn, is at the center of Clayton M. Christensen's disruption theory, first described in 1995.[9] According to the original framing, architectural inertia due to a company's links with existing customers prevented the company from responding effectively to "disruptive" change.[10] Twenty-five years later, this remains a fundamental tenet of the theory: that newer and smaller companies with fewer resources can challenge incumbents by addressing a neglected segment of the market.[11] At its core, disruption is still an outgrowth of architectural inertia. As inertia keeps the incumbent focused on existing customers (continuing what it has successfully done in the past), the entrant jumps in front of the incumbent by coming up with a novel solution.

Clearly, disruptive innovation is a critical—and popular—theme in strategy. But as Christensen and others have pointed out, it's often invoked to describe situations where it doesn't actually apply. Uber, for example, isn't really disrupting the traditional taxi business—it's *colliding* with it. Like Airbnb in the lodging industry, Uber meets recognized customer needs in a completely new (and highly threatening) way.

Collision, unlike disruption, involves more than introducing a technological innovation or revamping the business model or customer value proposition—it's about the emergence of an entirely different kind of company. As a result, defending oneself *against* collision can't be achieved by simply spinning off an online business, setting up a laboratory in Silicon Valley, or creating a digital business unit. It calls for rebuilding the core of the business and changing how the organization works, gathers and uses data, reacts to information, makes operating decisions, and executes operating tasks. Ultimately, it requires rebuilding the operating model, with software doing what many workers might have done in the past. This goes well beyond altering the patterns of human communication on which Conway focused.

Like Airbnb, Amazon, and YouTube, the companies that are driving collisions don't look or act like traditional companies. For better and for worse, they operate as software companies, fulfilling customer needs in new and more scalable ways. Furthermore, they are not constrained in any way by customary industry boundaries. They will use their universal capabilities in data, analytics, and AI, and their ability to generate network and learning effects, to increase their scope and the depth of interactions with their customers, causing collateral damage to those in their wake. Yet as they succeed by leveraging their scale, scope, and learning advantages, the digital operating models introduce a number of new problems. Among them: the preservation of privacy, algorithmic bias, cybersecurity, and increased market concentration.

As a new generation of players goes up against traditional companies, it is defining a new age and transforming our economy. The last time we saw changes of this magnitude was more than a century ago, with industrial leaders like GE, Sears, and Ford maintaining strong market positions for 50 to 100 years. New leaders are emerging today with very different operating structures. The way things are unfolding, the first dramatic effects of artificial intelligence will have less of an impact on human nature than on the nature of organizations, how they create and capture value, and how they shape the world around us.

9

The Future of Platforms

Michael A. Cusumano, David B. Yoffie, and Annabelle Gawer

The world's most valuable public companies and its first trillion-dollar businesses are built on digital platforms that bring together two or more market actors and grow through network effects. The top-ranked companies by market capitalization are Apple, Microsoft, Alphabet (Google's parent company), and Amazon. Facebook, Alibaba, and Tencent are not far behind. As of January 2020, these seven companies represented more than $6.3 trillion in market value, and all of them are platform businesses.[1]

Platforms are also remarkably popular among entrepreneurs and investors in private ventures. When we examined a 2017 list of more than 200 unicorns (startups with valuations of $1 billion or more), we estimated that 60% to 70% were platform businesses. At the time, these included companies such as Ant Financial (an affiliate of Alibaba), Uber, Didi Chuxing, Xiaomi, and Airbnb.[2]

But the path to success for a platform venture is by no means easy or guaranteed, nor is it completely different from that of companies with more-conventional business models. Why? Because, like all companies, platforms must ultimately perform better than their competitors. In addition, to survive long-term, platforms must also be politically and socially viable, or they risk being crushed by government regulation or social opposition, as well as potentially massive debt obligations. These observations are common sense, but amid all the hype over digital platforms—a phenomenon we sometimes call *platformania*—common sense hasn't always been so common.

We have been studying and working with platform businesses for more than 30 years. In 2015, we undertook a new round of research aimed at analyzing the evolution of platforms and their long-term performance versus that of conventional businesses. Our research confirmed that successful platforms yield a powerful competitive advantage with financial results to match. It also revealed that the nature of platforms is changing, as are the ecosystems and technologies that drive them, and the challenges and rules associated with managing a platform business.

Platforms are here to stay, but to build a successful, sustainable company around them, executives, entrepreneurs, and investors need to know the different types of platforms and their business models. They need to understand why some platforms generate sales growth and profits relatively easily, while others lose extraordinary sums of money. They need to anticipate the trends that will determine platform success versus failure in the coming years and the technologies that will spawn tomorrow's disruptive platform battlegrounds. We seek to address these needs in this article.

The Research

- This article and the book on which it is based, *The Business of Platforms*, build on some 30 years of research on the strategies and business models of platform companies.

- Using 20 years of data from the *Forbes* Global 2000, the authors identified the largest 43 publicly listed platforms built around the personal computer, internet services, or mobile devices from 1995 to 2015 and compared performance with a control sample of 100 nonplatform companies in the same set of businesses.
- Drawing on annual reports, the authors also identified 209 direct competitors to the 43 platform companies and analyzed reasons for the competitors' failures.
- Through interviews, case studies, and other sources, they identified common challenges faced by all types of platforms, as well as future trends for platform technologies and business models.

Platform Company Evolution

The companies that shaped the evolution of modern platform strategies and business models are familiar names. In the 1980s and early 1990s, Microsoft, Intel, and Apple, along with IBM, disrupted the vertically integrated mainframe computer industry. They made the personal computer into one of the first mass-market digital platforms, which resulted in separate industry layers for semiconductors, PC hardware, software operating systems, application software, sales, and services. A second wave of platform companies emerged in the mid-1990s, led by Amazon, Google, Netscape, and Yahoo! in the United States, Alibaba and Tencent in China, and Rakuten in Japan. They leveraged the internet to disrupt a variety of industries, including retail, travel, and publishing. In the next decade, social media businesses, pioneered by Friendster and Myspace, and then Facebook, LinkedIn, and Twitter, used platforms to enable new ways for people to interact, and for companies to target customers. More recently, Airbnb, Didi Chuxing, Grab, Uber, and smaller ventures such as Deliveroo and TaskRabbit have used platform strategies to launch the gig (or sharing) economy.

Today, platform companies are in nearly every market, and they all share common features. They use digital technology to create self-sustaining positive-feedback loops that potentially increase the value of their platforms with each new participant. They build ecosystems of third-party companies and individual contractors that allow them to bypass the traditional supply chains and labor pools required by traditional companies.

Moreover, all platform companies face the same four business challenges. They must choose the key "sides" of the platform (that is, identify which market participants they want to bring together, such as buyers and sellers or users and innovators). They must solve a chicken-or-egg problem to jump-start the network effects on which they depend. They must design a business model capable of generating revenues that exceed their costs. And finally, they must establish rules for using (and not abusing) the platform, as well as cultivating and governing the all-important ecosystem.

For all their similarities, it is possible to distinguish platforms on the basis of their principal activity. This yields two basic types: transaction and innovation platforms, with some hybrid companies that combine the two. (See "Basic Platform Types.")

- **Innovation platforms** facilitate the development of new, complementary products and services, such as PC or smartphone apps, that are built mostly by third-party companies without traditional supplier contracts. By *complementary*, we mean that these innovations add functionality or assets to the platform. This is the source of their network effects: The more complements there are or the higher quality they are, the more attractive the platform becomes to users and other potential market actors. Innovation platforms typically deliver and capture value by directly selling or renting a product, as in traditional businesses. If the platform is free, companies can monetize it by selling advertising or other ancillary services. Microsoft Windows, Google Android, Apple iOS, and Amazon Web Services are commonly used innovation platforms.

- **Transaction platforms** are intermediaries or online marketplaces that make it possible for participants to exchange goods and services or information. The more participants and functions available on a transaction platform, the more useful it becomes. These platforms create value by enabling exchanges that would not otherwise occur without the platform as an intermediary. They capture value by collecting transaction fees or charging for advertising. Google Search, Amazon Marketplace, Facebook, Tencent's WeChat, Alibaba's Taobao marketplace, Uber, and Airbnb are commonly used transaction platforms.

Hybrid companies contain both innovation and transaction platforms. Their strategies are novel because, in the early years of the PC and the internet, innovation and transaction platforms were distinct businesses. Connecting buyers and sellers, advertisers and consumers, or users of social networks appeared to be a fundamentally different activity from stimulating outside companies to create complementary innovations. In the past decade, however, a growing number of successful innovation platforms have integrated transaction platforms into their business models. Rather than lose control over distribution, the owners of these platforms have sought to manage the customer experience, like Apple has done with its App Store. Likewise,

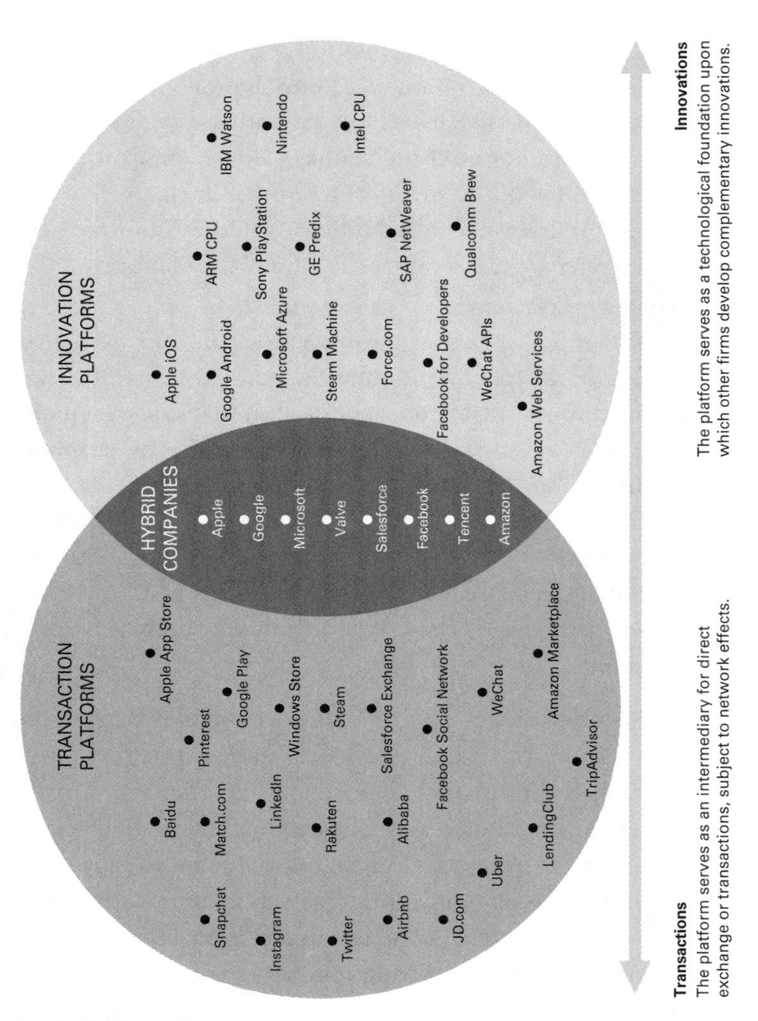

Basic Platform Types

In the quest for competitive advantage, companies are combining transaction and innovation platforms into a hybrid model. Source: *The Business of Platforms: Strategy in the Age of Digital Competition, Innovation, and Power* (Harper Business, 2019) some successful transaction platforms have opened their application programming interfaces (APIs) and encouraged third parties to create complementary apps and services. The owners of these platforms, such as Facebook and WeChat, recognize that not all innovation can or should be internal. Other prominent examples of hybrid strategies include Google's decision to buy Android, Amazon's decision to create multiple innovation platforms around Amazon Web Services and Alexa/Echo home AI devices, and Uber's and Airbnb's decisions to allow third-party companies to offer services that complement their ride-sharing and room-sharing platforms. Today, the most valuable global companies (which we mentioned above) all follow hybrid strategies.

Platform Company Value

Most platforms lose money (sometimes billions of dollars), but platforms that dominate their markets have been extraordinarily successful. When we compared the largest 43 publicly listed digital platform companies from 1995 to 2015 with a control sample of 100 nonplatform companies in the same set of businesses, we found that the two samples had roughly the same level of annual revenues (about $4.5 billion). But platform companies achieved their sales with half the number of employees. Moreover, platform companies were twice as profitable, were growing twice as fast, and were more than twice as valuable as their conventional counterparts.

In the process of examining the proxy statements and annual reports of the 43 success stories, we identified 209 platform companies that were their direct competitors but failed or disappeared as independent companies. The causes of these failures

Platform Business Performance, 1995–2015
An analysis of the performance of successful platform companies versus an industry control sample reveals the outsized advantage delivered by platforms.

Variable[a]	Industry Control Sample	Platform Companies
Number of Companies	100	43
Sales (millions)	$4,845	$4,335
Employees	19,000	9,872
Operating Profit %	12%	21%
Market Value (millions)	$8,243	$21,726
Market Value/Sales Multiple[b]	1.94	5.35
R&D/Sales	9%	13%
S&M + G&A/Sales[c]	17%	24%
Sales Growth Versus Prior Year	9%	18%
Market Value Growth	8%	14%
Total number of years of data for the sample firms	**1,018**	**374**

Source: *The Business of Platforms: Strategy in the Age of Digital Competition, Innovation, and Power* (Harper Business, 2019)

[a] Differences significant at $p < 0.001$ for industry sample versus platforms comparison using two-sample Wilcoxon rank sum (Mann–Whitney) test

[b] Market Value/Sales Multiple: ratio of market value compared with prior-year sales

[c] S&M + G&A/Sales: sales and marketing expenses plus general and administrative expenses divided by sales

were primarily mispricing (under- or overcharging) on one side of the market, oversubsidizing platform participants, or entering markets too late. The high number of platform failures supports the observation that even platform businesses can fail or struggle as the competitive environment or other factors change. For example, computing and communications platforms have faced continuous threats from new technologies over the past 40 years. Early success stories such as Myspace, Nokia, and BlackBerry saw their fortunes rapidly decline. Looking at the bigger picture, PCs cannibalized mainframes, smartphones cannibalized traditional cell phones, smartphones and the cloud are cannibalizing PCs, and so on.

In sum, platforms can become extraordinarily successful businesses, and some successful platform companies maintain their powerful positions for decades. However, the creation of a platform, even when it results in an IPO, is no guarantee of long-term success. The business must still be able to generate a profit and respond to change and competition.

Future Platform Trends

While the past 20 years have seen a dramatic expansion of platform-based technologies, applications, and business models, the next 20 years may see even more disruptive change. Digitization and emerging technologies such as artificial intelligence, machine learning, big data analytics, and infrastructure services have not yet attained their full disruptive potential. More and more individual user and transactional data will become connected with different platform services and functions, with the potential to generate positive and negative outcomes.

No one can predict the future, but we have identified four major trends that are likely to affect platform dynamics across industries: the emergence of the hybrid model as the dominant strategy for platform businesses, the use of AI and machine learning to produce major improvements in platform operations and capabilities, increasing market concentration by a small number of powerful platform companies, and the demand for more platform curation and regulation to address problems unleashed by some of today's platform companies.

Trend 1: More hybrid business models. Competition and the potential of digital technology and data will turn more and more platform companies into hybrids. The underlying driver of this trend is digital competition. Unlike in the traditional economy, where companies require expensive physical investments to build out their business models, in the digital world, companies can grow rapidly with a clever combination of data, software, and ecosystem strategies.

Trend 2: More turbocharged innovation. Next-generation platforms will drive innovation to a new level. Advances in artificial intelligence, machine learning, and big data analytics are already enabling organizations to do more things with less investment, including building businesses that were impossible in years past. Although AI is still in its nascent phase, Google, Amazon, Apple, Microsoft, IBM, and other companies are no longer treating the technology as fully proprietary. Instead, they have turned some of their AI capabilities into platform services that third parties can access and build upon for their own applications. The combination of platforms enabling the capture of more data, with the ongoing improvements in cloud computing, should allow future platforms to enable a wide range of new applications, such as products with voice interfaces and driverless cars.

Trend 3: More industry concentration. The total number of platforms has been exploding, and dominant market shares, as well as strong network effects, have been increasingly difficult to attain because of *multihoming* (the ability of platform users and complementors to access more than one platform for the same purpose, such as using both Lyft and Uber for ride-sharing).

In the years ahead, virtually all large platform companies will evolve from free marketplaces to curated businesses with increasing government oversight and potentially new types of regulation.

Nevertheless, in coming years, we expect to see even more market power concentrated in a smaller number of large platform companies.

This paradoxical situation will result because some markets will tip toward one platform and further concentrate market power. Witness IBM's ascension to the pinnacle of platform power in the computer industry in the 1960s and 1970s, and Intel's and Microsoft's in the 1980s and 1990s. In the past decade, the number of markets that appear to have tipped to a few dominant players has expanded, with Amazon, Alibaba, Apple, Google, Facebook, Microsoft, Tencent, and Uber, among others, achieving market shares well over 50%.

Trend 4: More curation and regulation. Mark Zuckerberg based his early dictum to "move fast and break things" on the premise that good things will happen if we connect the people of the world. Most platform entrepreneurs and investors agreed with him: They believed that platforms would connect people with products and services at ever-decreasing prices and free the world from the frictions and imperfections of traditional and local marketplaces. As it turns out, not all actors in the digital world are do-gooders. Those engaged in partisan politics, spies, terrorists, counterfeiters, money launderers, and drug dealers all found ways to use digital platforms to their advantage.

Once the platforms reach a scale at which they can affect social, political, and economic systems, their owners increasingly need to evolve from hands-off to hands-on curation. In the years ahead, virtually all large platform companies will evolve from free marketplaces to curated businesses with increasing government oversight and potentially new types of regulation. Although it is a cliché, for the world's biggest platforms, growing power means increased responsibility—and oversight.

Three Emerging Platform Battlegrounds

Several competitions are currently underway that illustrate the trends above and offer insight into what might come next in platform technology and strategy. Several fast-emerging fields—AI, cloud computing, and, ultimately, quantum computing—will enable disruptive innovations as well as changes in business models.

Voice wars: Rapid growth, but chaotic competition. Recent advances in machine learning and the subfield of deep learning have led to dramatic improvements in pattern recognition, especially for images and voice. Apple got the world excited about a voice interface with the introduction of Siri in 2011. For the first time, consumers had access to a natural conversation technology that worked (at least some of the time). Despite its first-mover advantage, however, Apple's strategy for Siri was classic Apple: It designed Siri as a product to complement the iPhone, not as a *platform* that could generate powerful network effects in its own right.

Enter Amazon. When it introduced the Echo speaker and Alexa software in late 2014, it set in motion a war for platform domination among Alibaba, Apple, Google, Microsoft, Tencent, and a host of voice startups. Amazon's strategy was to link multiple platforms powered by Amazon Web Services and offer a combination of speech recognition and high-quality speech synthesis with various applications. Immediately identifying the potential for network effects, Amazon launched its Alexa Skills Kit—a collection of self-service APIs and tools that made it easy for third-party developers to create new Alexa apps. This open-platform strategy accelerated the number of Alexa skills from roughly 5,000 in late 2016 to more than 90,000 in 2019.

Amazon's success spurred Apple, Google, Samsung, and various Chinese companies into action. By late 2017, voice had morphed into a classic platform battle: Amazon and Google began heavily discounting products to build their installed base, with each side racing to add applications and functions. All the major players have also been licensing their technologies (often for free) to consumer electronics, automotive, and enterprise software companies, hoping that these companies will use their voice platforms and solutions.

How the platform war in voice computing will evolve depends heavily on the ease of multihoming. Currently, consumers can easily switch voice platforms or use more than one. It will also depend on how the players choose to position themselves. There are many opportunities for competitor differentiation and niche competition in voice: Apple has focused on the quality of music, Amazon on media and e-commerce, and Google on search-related queries, to name only a few.

Meanwhile, competitive advantage has not yet hardened into market concentration. Google has already embedded its voice capabilities into hundreds of millions of Android devices. But Amazon has the largest smart-speaker installed base, with tens of millions of devices sitting in users' homes, especially in the United States.

Ultimately, we expect the winner or winners in voice to be those platforms that build the largest installed base of users *and* create the more vibrant ecosystems for producing innovative applications. These ecosystems are likely to generate compelling voice solutions that reduce platform multihoming and competition from niche players and differentiated competitors.

Ride-sharing and self-driving cars: From platform to service. While AI will spawn a range of new products, platforms, and services, it will also enable new capabilities that create, enhance, and destroy existing businesses. Nowhere is this dynamic clearer than in the emergence of self-driving cars, where Japan's SoftBank has invested $60 billion in 40 companies, including Didi, Grab, and Uber. Although Uber has already fallen far below its peak valuations, and other investments may follow, SoftBank is betting that transportation services platforms, such as ride-sharing accessed through smartphones, will eventually become highly concentrated businesses, generating huge returns similar to Alibaba, Apple, Google, and other digital platforms.[3]

Ironically, this new AI-powered technology not only threatens the century-long hegemony of automakers but may also disrupt today's ride-sharing platforms. Despite relatively strong network effects between users and drivers, innovation in technology and business models could replace the platforms belonging to companies such as Didi, Grab, Lyft, and Uber.

The business challenge for ride-sharing platforms is simple: They tend to lose money, and lots of it. Unlike asset-light transaction platforms such as eBay, Expedia, or Priceline, ride-sharing platforms are not fully digital businesses: The ordering and payment transaction is digital, but the service delivery is physical, with mostly local and limited economies of scale and scope. Furthermore, the cost of attracting and paying drivers while keeping fares below the market price of taxis has squeezed the profit potential and resulted in huge losses for these companies. In addition, many drivers and riders multihome: They drive for or use both Uber and Lyft, as well as conventional taxis.

The bottom line is that *platformizing* a low-margin business like taxi services or food delivery does not necessarily make it a profitable business, like selling software products or other digital goods. As a result, Didi, Grab, Lyft, and Uber have announced that their long-term strategies are to move beyond purely transactional platforms that match riders with drivers to transportation as a service. As Lyft CEO Logan Green said, "We are going to move the entire [car] industry from one based on ownership to one based on subscription."[4] In this new model, ride-sharing platforms will probably own or lease fleets of automobiles, as well as bicycles and scooters.

Tech companies like Google and most of the major automobile manufacturers, including General Motors and Toyota, are also investing aggressively in similar strategies. Despite a long history of selling products, even the most conservative car companies see AI as a way to transform themselves into service companies.

Autonomous vehicle technology promises to remove human drivers, which would dramatically drive down the marginal cost of transportation services for ride-sharing platform owners. But, in addition to bringing new competitors into the industry, it would also require massive capital investments in R&D and fleet costs. Some observers see this combination of conditions forcing Uber and other ride-sharing platforms to "either figure out a way to buy or at least manage an enormous fleet . . . or face annihilation by others who will."[5] In response to this threat, Uber began investing in autonomous vehicle technology in 2014. Lyft has taken a different approach by trying to form partnerships through its Open Platform Initiative.

Owning or leasing a fleet of autonomous vehicles is counter to the two-sided platform business model of matching riders with drivers and their cars. If they make the transition to autonomous fleets, Uber and Lyft will become one-sided, company-controlled platforms that own and resell their own assets. The risk is that self-driving car services are unlikely to materialize as

Ultimately, we expect the winner or winners in voice to be those platforms that build the largest installed base of users *and* create the more vibrant ecosystems for producing innovative applications.

quickly or be as profitable as purely digital platforms with high transaction volumes. Nonetheless, future consumers are likely to benefit from more and cheaper ride-sharing services, as long as these businesses have enough capital and cash flow to survive.

Quantum computers: A next-generation computing platform. In 1981, Nobel laureate Richard Feynman challenged his fellow scientists to build a computer mimicking nature—a quantum computer. The challenge was accepted. In 2015, McKinsey consultants estimated that 7,000 researchers were working on quantum computing, with a combined budget of $1.5 billion.[6] By 2018, dozens of universities, approximately 30 major companies, and more than a dozen startups had notable quantum computing R&D efforts underway.[7] More recently still, Google announced that it had built a quantum computer that far exceeded the capabilities of the world's fastest supercomputers, at least for specific types of calculations.[8]

The state of quantum technology today resembles that of conventional computing in the late 1940s and early 1950s: Quantum computers are difficult and expensive to build and program, and reside primarily in universities and corporate research labs. Nonetheless, they represent a revolutionary innovation platform, with the additional potential to stimulate new transaction platforms for specialized applications in simulation, optimization, cryptography, and secure communication.

Will quantum computing produce successful new platform businesses? Currently, the network effects appear weak because the application ecosystems are still nascent and divided among several platform contenders. A spin-off from the University of British Columbia named D-Wave Systems, founded in 1999, has the lead in applications and the largest patent portfolio, followed by IBM and Microsoft. However, D-Wave has not built a general-purpose quantum computer, unlike most other entrants into the field, and recently IBM has taken the lead in annual patent filings. To build better programming tools and test real-world applications, more researchers must gain access to these patents and to more-powerful quantum computers.

Quantum computers will not replace digital computers. Nor do we see this field as a winner-takes-all-or-most market in which one company's unique architecture will dominate, as occurred in mainframes, PCs, smartphones, microprocessors, consumer electronics, and other markets. Quantum computers will most likely always be special-purpose devices for certain types of massively parallel computations, with different technologies more useful for particular applications.

At the same time, quantum computing platforms are likely to face intense scrutiny and regulation because of the potential cryptography applications. On the one hand, quantum computers may be able to break secure keys generated by the most powerful conventional computers, which now protect much of the world's information and financial assets. On the other hand, quantum computers themselves could generate unbreakable keys and facilitate truly secure communication. The leading companies will have to regulate themselves as well as work closely with governments, which are likely to play a major role in overseeing some of these new applications and services.

Platforms as Disrupters

We are heading into a future where we will buy and own fewer products (cars, bikes, vacation homes, household tools, and so on), and we will contract for more services directly with one another. We will likely manage this sharing through peer-to-peer

Massive infusions of capital are a third form of disruption that could be just as powerful as new technologies and business models.

transaction platforms along with general-purpose digital technologies, such as blockchain, to enable more secure and transparent exchanges.

Some platforms that enable this future will follow the model of disruption that Clayton M. Christensen described, with cheaper, inferior technologies gradually overtaking incumbents. This occurred with the gradual domination of personal computers over mainframe computers and the rise of e-commerce and internet marketplaces over traditional stores, though the older technologies and ways of doing business continue to exist. We expect to see similar Christensen-style disruptions in the future, with voice platforms and self-driving cars.

But this is not the only type of disruption we expect to see in the platform economy. Our research illustrates how platform disruption can come from above, as well as from below. For example, Apple and the iPhone disrupted the smartphone industry by building a high-end platform with superior performance and features from the very beginning. Similarly, quantum computing systems and applications such as cryptography or complex simulations will likely arrive as expensive solutions coming from the high end of the market.

Massive infusions of capital are a third form of disruption that could be just as powerful as new technologies and business models, such as turning transportation into a subscription service. The use of smartphones to match drivers and riders was innovative as a business model and required only modest investments in new technology. But what is less remarked on is the fact that Uber and other ride-sharing platforms disrupted the taxi business by spending billions of dollars in venture capital to subsidize a low-margin commodity transportation business. Whether or not Uber and similar ventures survive, and whether or not financial backers such as SoftBank ever recoup their investments, they have disrupted the taxi business forever.

In short, industrywide platforms and their global ecosystems have already disrupted many aspects of our personal and working lives. New innovation and transaction platforms have enabled nearly every type of exchange and activity imaginable in today's world, and platform entrepreneurs have made Anything-as-a-Service possible. No matter how they evolve, we expect that future platforms will continue to inspire both innovation and disruption.

10

The Experience Disrupters

Brian Halligan

Let me tell you about my evening routine. Every night, my dog Romeo and I come home from the Cambridge, Massachusetts, offices of HubSpot, where I'm CEO, by taking a Lyft. We play our favorite band on Spotify. Cranking the music, we boogie over to the dog area, clean out some of Romeo's toys, and see if he got a new package in the mail from Chewy—he loves their chicken lollipops. After a snack, I head down to the gym for a workout I booked through ClassPass. I come home and shower and shave using a new package from Dollar Shave Club. I order something from DoorDash, and, after it arrives, Romeo and I put our toes up and check out a favorite movie on Netflix. Then we lie down on our Casper mattress, and we get a good night's sleep.

I think we have a fascinating evening routine. Why? Because all these companies—I just ripped through eight of them—have replaced companies I used to do business with.

It's not just my evening routine; it's my daily routine. It's all of our daily routines, isn't it? There's been a massive wave of disruption happening in the consumer world, courtesy of companies like Lyft, Netflix, and Spotify.

The same shift is going on in the business world. When I'm on the West Coast, I set up in a remote office and collaborate with team members on Slack. When there's a meeting, I fire up Zoom. When I'm hungry, I scarf down something from ezCater. Again, this is a wholesale swap of vendors.

But this isn't disruption in the way most of us think of it. We tend to think about *technology* disrupters—the browser, Google, Intel, the iPhone, maybe the Tesla someday. Big technology companies with lots of patents. (In 2018, Intel was granted 2,735 patents, Apple 2,160, and Google 2,070.)[1]

Companies like Chewy and Dollar Shave and ClassPass—are they technology disrupters? I'm not so sure. I went very deep on this list of companies plus a few others, about 20 altogether, with two of my colleagues at HubSpot. We talked to almost all of these companies' founders. We purchased pretty much all their products, we read all their terms and conditions, we talked to their big investors. We asked about their patents and found only about 50 total. And my theory is that these companies are not technology disrupters.

Instead, I think we're seeing a new species of disrupter emerging in our economy, a species I call *experience disrupters*. These organizations all have great products, but they offer even better experiences. How they sell is why they win.

Of course, all companies aim (or should aim) for great customer service, but that's not what I'm talking about here. These companies have fundamentally reshaped what their customers come to expect in the experience of purchasing and using their product or service. This is a central insight of Clayton M. Christensen's Theory of Jobs to Be Done, which tells us that customers don't simply buy products or services. They hire them to do a job for them. Doing that job well for customers involves creating the right experiences for those customers, from the moment they begin to think about purchasing the product to their everyday use of that product. It's an essential part of developing a deep relationship with customers: You solve their struggle for them.

What I think we are seeing now are companies that outmaneuver the competition by excelling at this. After studying such companies, what they're good at, and the customer experience with each of them, I've come up with five things I call *modern adaptations* that allow these experience disrupters to run over the incumbents in their industries. Here, I'll discuss what I've observed about those adaptations, leaving you with a playbook to use in your company.

They Give You Experiences You Didn't Know You Wanted

The first adaptation is that while incumbent companies focus on *product-market fit,* experience disrupters work on *experience-market fit.* Product-market fit, when you've found the right mix of product for just the right target market, is considered by these companies as necessary but insufficient to get the disruption they're really after. For experience disrupters, what matters is offering *experiences* that surround the product and that customers didn't even know they wanted or could ask for.

Let me give you an example. I first heard of Carvana when we started this research project. It turns out a bunch of my colleagues had purchased cars from this online used-car company and were raving about it. Carvana was founded in 2012 and was the eighth-largest used-car dealer in the United States in 2018.[2] It went public in 2017. As of this writing, it has a market cap of roughly $12.5 billion. And it is a killer experience disrupter.

How did Carvana become so successful so fast? You might think it was about inventory: Typically, a car dealer has cars all

While incumbent companies focus on *product-market fit*, experience disrupters work on *experience-market fit*.

over parking lots, and Carvana instead has a giant online car vending machine.

Now, that step is necessary, but insufficient, to get the crazy growth it's had.

The reason Carvana has exploded is that it's focused on the experience-market fit. The company's leaders set out to create a whole new way to buy a car. You have a very Amazon-like experience, in the sense of how user-friendly the online interface is. You choose the price range, mileage, condition, and type of car you want. You can get alerted when a car in your range is available near you. Once you select a car, you can view a 360-degree inspection with annotated zoom-in areas to see where there is wear and tear.

But you don't just buy the car from Carvana: The company deals with the department of motor vehicles, it deals with the taxes, it deals with the registration. It does all the crapola that none of us wants to do. Then you tell Carvana, "Hey, I bought the car, and I want it delivered to my house on Tuesday afternoon"—you pick a time and a place, and the company brings it to you. Awesome. And *then* you drive the car around for a week, and if you're not happy with the car for whatever reason, you can return it, no questions asked.

Carvana has taken the cringeworthy process of buying a car and automated it, institutionalized it, and made it awesome. *That's* experience-market fit.

They Make Interactions Frictionless

The second adaptation is that experience disrupters pull the friction out of each customer interaction. The analogy I like is the mechanical flywheel—the circular device that can provide a continuous power output. In this analogy, the less friction customer interactions have, the faster the flywheel spins, and the faster a company grows. In businesses that are struggling to keep up with experience disrupters, their flywheels are full of friction. Experience disrupters are very good at reducing that tension.

Consider Atlassian, an Australian B2B collaboration software company that is a friction-fighting superhero. It's a large company that is growing very fast and is very profitable, with a market cap near $36 billion. The company's president, Jay Simons, serves on HubSpot's board, and we are one of Atlassian's biggest customers, so I know the business well. Simons told us that changing the process of buying B2B software meant rethinking how the marketing and sales departments interact with customers, and even how the contracting process works.

First, Atlassian's marketing department looks just like a B2C marketing department, focusing less on generating new leads and more on activating current users and multiplying the number of users and teams within a customer. Instead of fighting the uphill battle for senior-level evaluation of their solution, Atlassian focuses on the ease with which an end user can invite a colleague to a collaborative project.

Now, most of the B2B experience disrupters do a really nice job of marrying low-friction, B2C-style marketing with a slightly-heavier-friction traditional enterprise sales model. But Atlassian doesn't do this. What impresses me is that most of its transactions happen without the sales team. Salespeople negotiate the highest sticker price deals—basically, the deals that generate the top 1% of value. Otherwise, sales are straightforward, with no commissions. Just a few years ago, you'd buy a toothbrush or a comb online, but now people are buying multimillion-dollar pieces of software the same way.

Atlassian also tweaked the contracting process. Think about how the process typically works: Potential buyers will Google something they need, find a product on a website, and possibly check out the company's blog and social media and do the same with its competitors. They'll call the company, ask to talk to someone on the sales team, and maybe have a great experience with a salesperson who engages them, understands their pain, and solution-sells them.

Trust and goodwill are built up, and the customer is ready to buy. And then: A brutal negotiation over the course of weeks or months ensues.

All that trust, all that goodwill, all that goodness—it goes down the tubes. I really don't like this. Jay Simons doesn't like this, either. So what he said was, "*Basta*. Enough. No more negotiations. I'm not giving discounts to anyone. I don't care if it's my sister. No discounts." What he wants to do is keep the goodwill between us. He doesn't want an adversarial relationship. So when a prospect asks the inevitable question, "How about a discount?" Atlassian staff are trained to explain that the software is relatively lower cost because the company builds discounts directly into the prices to treat every customer equally and to take away price uncertainty. The purchase price is online, and because they don't negotiate changes in prices or terms and conditions, the contracting process is not complex—and it's easily automated. All these decisions eliminate friction at this stage of the sale.

They Personalize the Relationship

The third adaptation is that experience disrupters are awfully good at creating a personalized experience. Their competitors, the incumbents in the industry, offer a more generic experience when they're prospecting customers. In our research project, when we talked to the founders of experience disrupters, I was surprised at how much they didn't sound like tech people. The language they used made them sound more like executives from The Ritz-Carlton or the Four Seasons. The way these companies cater to each customer makes them less like tech companies than like ultramodern hospitality companies.

Think about Netflix. Inside the company's database, there's a fingerprint for every one of us customers. The more we use their product, the more shows we watch or click on or give up on 10 minutes in, the better the company gets at personalizing its recommendations to us. Netflix suggests new content based on viewing history, but even the finest details—such as the thumbnails that accompany each show—are tailored to an individual user's browsing habits. This is one of Netflix's real secrets of success.

Now, Netflix isn't the only company using data to be much more prescriptive about experience. This is also happening at Stitch Fix, an online personal styling company based in San Francisco. The company offers customized clothing selection for customers and also sells the outfits. When Stitch Fix first got started, individual stylists recommended combinations of apparel solely on the basis of lengthy profiles completed by customers about their style preferences and specific measurements.

But Stitch Fix CEO Katrina Lake knew the value of data to deepen the accuracy of stylists' recommendations and to give scale to the business. Today, in addition to the initial customer profile, the company uses direct feedback from customers on their purchases, mountains of data from across all its customers about which items were purchased together and which were rejected and returned, and fastidious details from its merchandise team about the precise measurements, textures, and aesthetics of each clothing option. All this arms Stitch Fix with an opportunity to base recommendations on much more than just "customers who bought this also bought that" logic.[3]

The company's algorithm helps generate recommendations that have progressively led to increased purchases over returns, and more additional purchases by repeat customers. It's working: Stitch Fix, which went public in 2017, has a market cap of $2.4 billion.

Netflix and Stitch Fix are playing the same game: They use lots and lots of data to highly personalize your experience with them. How they sell is why they win.

They Get Customers to Sell for Them

The fourth adaptation is that while the incumbents know how to sell *to* their customers, the experience disrupters are very good at selling *through* their customers. One of my favorite examples is Emily Weiss, founder of the cosmetics company Glossier. She started off as a blogger—she's a fabulous content creator, and her blog, *Into the Gloss*, was blowing up with beauty tips. And then she started developing beauty products.

Where Weiss is next-level and a bona fide experience disrupter is her ability to not just create her own content but also encourage and enable her customers to create content. Glossier makes its products available to popular video bloggers, known as vloggers, sometimes even prior to public release to build buzz. For instance, Glossier worked with Jackie Aina, a Top 20 YouTube beauty vlogger who has more than 3 million YouTube subscribers, to review a product when it was still unannounced.

Thousands of wannabes and micro influencers who may have a few thousand followers each imitate the most popular vloggers with their own video reviews. The result is hundreds of thousands of pieces of content out there about Weiss's products—all created by her customers. Some of those individual videos have more than a million views. Glossier is still a private company, but its estimated valuation is $1.2 billion.

Warby Parker, the eyeglasses company, is another classic experience disrupter for similar reasons. Neil Blumenthal, the cofounder and co-CEO, thought the old process of buying glasses was a pain. And it was. You had to schedule going down to the store, and the scheduling was a bear because you had to bring your most judgy friend with you. Blumenthal said, "I'm going to rethink that. I'm going to mail you the glasses so you try them on, you can post photos on Instagram, and you can then ask *all* your judgy friends which one they like." Again: How they sell is why they win.

They Empower Employees to Make Things Right for Customers

This brings us to the fifth adaption: Experience disrupters enable customer-facing employees to fix things when they need to.

Traditionally, companies woo customers to make a purchase, but the second that purchase is made, it becomes the customer's hassle to get service on it or return or exchange it if there's a problem. Lots of companies offer free shipping, for example, but customers have to pay for the shipping to make a return, they have to have kept the receipts, and they have to pay attention to how long ago the purchase was made.

Experience disrupters make all these details much more customer-friendly. I was surprised at how powerful this play was. By rethinking something as mundane as terms and conditions, they are able to bust through those their industry models in effective ways.

A great example is online pet store Chewy, which I mentioned earlier. I ordered a medium shirt from there for Romeo. He's always taken a medium. But when I put the shirt on that poor dog, he could barely breathe. It was too tight—too many of those chicken lollipops. So I called Chewy, and I said, "I'd like to return my medium for a large," and the woman said, "Nope, that's not how we're going to do it today." She said, "Give your medium to a friend of yours, and we'll send you a large for free."

Chewy gives its customer service reps a discretionary budget to create opportunities to build goodwill with customers, and this empowerment allows for a customer experience that feels seamless. Obviously, this worked out great for me: I didn't have to do the return, I didn't have to do any paperwork, and Romeo got a shirt that fit. And it worked out really well for his friend, Woodford, who now has a new free shirt.

What I like about this model is that Chewy's costs to acquire Woodford as a future customer were very low, right? I got Woodford for them. Chewy didn't have to spend very much to do it. And the total lifetime value of Romeo is now very high.

Experience disrupters know how incredibly significant it feels for customers when there's a genuine change in the power balance in post-sale interactions. A friend of mine got herself all worked up before calling her cell phone company about something she thought was a mischarge on her bill—a situation most of us have been through with a phone company or cable

Embrace these five experience disrupter plays and harness your inner superhero.

provider—anticipating yet another interaction that would go badly. She hung up in near disbelief just a few minutes later when the customer service rep believed her, fixed the billing charge while they were on the phone, and offered her a courtesy credit for the hassle. There is unbelievable value in this adaptation.

These experience disrupters really are a different species. They think differently, and the founders have a healthy disdain for conventional wisdom. They spend hardly any of their energy *extracting value* from their customers. Instead, they spend all their energy thinking, "How do I *add value* for my customers?" They're really good at this stuff. One last time: How they sell is why they win.

Here's a summary of the five points:

- Don't obsess completely about product-market fit. Obsess about experience-market fit. Embrace your inner Carvana.
- Remember that dollars flow where the friction is low. Mechanically remove friction. Automate like the superheroes at Atlassian.
- Personalize, personalize, personalize. Stop embracing automation without personalization—that's what people call spam. Think like Netflix. Dust for fingerprints.
- Sell *through* your customers, not just to them. Let Glossier be your model.
- Rethink how customers get treated after the sale. Look at your terms and conditions. Give your customer-facing employees the tools to make things right. Delight people, the way Chewy does.

I started this article talking about my nightly routine. Routines can be good. But they can also hold you back.

You have routines in your job, and when you finish reading this, you're going to return to whatever it is that you do. You have a choice. You can do your normal routine: Drag the spreadsheet on your career, drag the spreadsheet on your company. Or you can set out on a new course, a more exciting one. Stop meeting your customer needs, and start exceeding them. Choose to become an experience disrupter.

11

The New Disrupters

Rita Gunther McGrath

Clayton M. Christensen's Theory of Disruptive Innovation first came to public attention 25 years ago. Christensen presciently explained that fast-moving disrupters entering the market with cheap, low-quality goods could undermine companies wed to prevailing beliefs about competitive advantage. In the last decade, however, the profile of disrupters has changed dramatically. The critical difference is that they now enter the market with products and services that are every bit as good as those offered by legacy companies. Their ascendance doesn't undermine Christensen's theory. In fact, they expand its reach and vitality—and make it harder than ever for traditional companies to compete.

The Classic Theory of Disruption

Before we look at how things have evolved, let's briefly review why Christensen's theory proved so influential and, indeed, disruptive to existing ideas of competitive advantage.[1] Traditional strategy had been anchored on the notion of "generic strategies" in which a company could compete at the high end by differentiating, at the low end by pursuing cost leadership, or focus on serving a specific niche exceptionally well.[2] Christensen illustrated a way for new entrants to cheerfully ignore these basic strategy dynamics. He showed how a new kind of dangerous competitor could wreak havoc by entering at the low end of a market, where margins are thin and customers are reluctant to pay for anything they don't need.

The new entrant comes in with a product or service that's cheaper and more convenient but that doesn't offer the same level of performance on the dominant criteria that most customers expect from incumbents that have been working on the technology for years. The incumbents feel they can ignore the newcomer. Not only are its products inferior, but its margins are lower and its customers less loyal. Incumbents choose instead to focus on sustaining innovation—making improvements to the features that have been of most value to their high-end customers.

Christensen showed the downside of ignoring the newcomers. Eventually, as these upstarts improve, they become pretty good at the old dominant criteria. They also develop such solid innovations at the low end that they bring new customers into the market. Having doubled down on what has always worked, the incumbents fail to notice two things. First, they miss out on the meaningful value of the low-end innovations developed by newcomers. Second, they are late to recognize that their own customers are less willing to pay more for more of the old attributes. Their key product has been commoditized, supplanted by a new technology that better suits the changed needs of customers.

A prime example of this process occurred at Intel. The chipmaker enjoyed decades of high margins by selling high-end, powerful, and fast computer chips for laptops, desktop computers, and servers that allowed users to get the most out of increasingly power-hungry software. The company (and its customers) didn't care much about power consumption because personal computers were either permanently plugged into a power source or had sufficiently large batteries to go hours between charges. Dominant to the point of near-monopoly, Intel dismissed and largely ignored a new set of less powerful, albeit less power-hungry, chips based on the ARM architecture (created by a once-obscure British company).

The smartphone revolution of the late 2000s exposed the fatal flaw in Intel's offerings. The company's chips were power-hungry, but now users wanted light mobile devices that could last all day. Chips based on the ARM design were far more efficient—the new differentiating quality. Intel managers had been focused on making its core microprocessors better at what had always seemed to matter most. So the company missed the potential of mobile device chips, which more than made up for their lower margins by finding their way into billions—not millions—of devices.

Intel's struggles with chips for mobile devices illustrate two dimensions of the disruption described by Christensen. The first is the market entry of a new competitor whose offerings are not good enough to meet the needs of established customers (PC owners). The second is the moment when that entrant creates a market by selling solutions to users who were never customers before, like smartphone manufacturers.[3]

Christensen's theory also highlighted the powerful way that management metrics and incentive structures reinforce this pattern. In his view, many of these combine to discourage executives from investing in innovation. Financials expressed as ratios, accounting-driven depreciation schedules, conventional business plans, and stock- or time-based rewards to managers all detract from a leader's willingness to pursue uncertain (though potentially high-payoff) innovations. This has all been exacerbated by outsize rewards to executives and investors in the short run, which undermine investment for the long run.

The Rise of the Cheap, Convenient, and *High-Quality* Startup

Today's direct-to-consumer (DTC) disrupters illustrate a next evolution in the theory of disruption. These disrupters target the very core of incumbents' existing businesses by using today's broad array of powerful digital technologies to offer products or services that are cheaper, more convenient, and every bit as good as existing offerings. The combination of "just as good" with new digital technologies creates a massive inflection point, putting more pressure on incumbents than ever.

The traditional assumption underpinning most retail businesses was that the gatekeeper for product sales would be large distributors like Walmart, Target, or Carrefour. Vendors had to demonstrate large enough demand for a big enough customer segment to earn space on a distributor's shelves. This gatekeeping function created several side effects. The first was that products themselves needed to be fairly standardized to offer consistency across the many distributor locations. The second was that the manufacturers had relatively crude information about who the buyers were, how the products were stocked and displayed, and how they performed relative to competitors sold by the same vendor. The producer had relatively little control over the buying experience. For many producers, the most important "customer" was the retailer, not the end user.

The 2010s saw an explosion of consumer-products companies that dispensed with the gatekeepers to offer their products directly to consumers, hence the moniker *D2C*, or direct to consumer. Companies such as Warby Parker (founded in 2010), Dollar Shave Club (2011), Glossier (2010), Away (2015), Casper (2014), and Bonobos (2007) are upending categories as varied as eyeglasses, men's grooming, skin care, travel, mattresses, and clothing. Their value proposition to customers almost always features a comparable product at a lower price. More important, it always offers a shopping experience that eliminates many of the frictions and irritants of conventional retail.

Dollar Shave Club (acquired by Unilever for $1 billion in 2016) made a point of criticizing flaws in the existing business model of conventional men's shaving products. In a hilarious video that went viral on social channels, cofounder Michael Dubin mocked the incumbent's practices. "Do you like spending $20 a month on brand-name razors?" he asks his mostly youthful audience. "Nineteen goes to Roger Federer! . . . Stop paying for shave tech you don't need." Gillette, the incumbent, was forced to react by reducing prices, launching a shave club of its own, and even venturing into edgy advertising (to mixed reviews). Still, its market share has suffered, both from the competition with D2C companies like Dollar Shave Club and Harry's and from the trend among men, particularly younger ones, to wear beards.

In a short period of time, new competitors have radically changed customer behavior in three significant ways:

- **Consumers are now happy to purchase hard goods like mattresses, furniture, and even cars online.** Previous generations found it unthinkable to do so. Making a wrong choice could involve expensive returns, lots of wasted time, ongoing quality and safety worries, and potentially even financial losses. Since making a wrong choice was seen as risky, the assumption was that consumers would always want to touch and feel such goods before making a big-commitment purchase. The new disrupters have eliminated that risk and complexity. Don't like your Casper mattress after 99 days? No problem—the company will come pick it up from you, free of charge. Casper has reduced the risk of making what once was a high-stakes decision. In 2019, the company's revenues topped $500 million, taking a chunk out of an industry whose sales were reportedly $27 billion the same year.

- **Almost everything can be sold as a service.** While the idea started with software (the famous software-as-a-service, or SaaS, model), you can now utilize clothing, furniture, cars, trucks, heavy equipment, and even pet supplies on a subscription or limited-trial basis. Why bother to own products when you can get the same benefits only as needed, with flexible spending?
- **Excess capacity is a consumer asset.** The poster child for this trend is Airbnb, which created a marketplace in which ordinary people with underutilized real estate could make money by renting their space to strangers. As of 2019, Americans reportedly spent more money with Airbnb than they did with Hilton, accounting for some 20% of consumer money spent on lodging. The model is now extending to other asset classes, with startups like Neighbor.com, which connects homeowners with excess room at home to people who need storage space.

The new disrupters have eliminated risk and complexity. Don't like your Casper mattress after 99 days? No problem – the company will come pick it up from you, free of charge.

Given the success, reliability, and proven value of these new D2C competitors, it's hardly surprising that the value of many established brands is in sharp decline. Just as digital technologies allow companies to build businesses almost overnight, social media, digital channels, and online influencers can help new brands build meaningful identities and reputations at warp speed.

The Digital Elements of the New Disruptive Model

These new D2C businesses have several similarities, each driven by digital technologies, algorithms, data analytics, and new forms of connectivity.

- **Access to assets, not ownership of assets.** Traditional organizations used the assets they owned to both create competitive differentiation and establish entry barriers. The D2C organizations, instead, participate in digital platforms that can virtually represent both sides of an on-demand transaction, removing friction and risk. Contracting for asset usage on an open market allows them to scale quickly. However, it also makes their business models relatively easy for others to copy. The online mattress-in-a-box business, for instance, is thought to have as many as 150 new entrants. The amount of new entry echoes what Harvard professors William Sahlman and Howard Stevenson years ago called *capital market myopia,* in which startups charge into a category that can't possibly sustain all of them.[4]
- **Cocreation with customers.** Digital channels eliminate middlemen. As their name implies, D2C companies create a direct relationship with their customers. This gives them powerful feedback loops in which they can more rapidly experiment,

iterate, and customize offerings with far more flexibility than a traditional retailer. The best D2C brands create a complete end-to-end experience, capturing the customer's attention, loyalty, and data through the entire process rather than sharing it with anyone else.

- **Always-on and mobile.** There have always been organizations that sold directly to consumers (think L.L. Bean or Lands' End). The new breed of D2C companies, however, uses mobile technology and mobile infrastructure to make interaction a 24-hour, always-on experience. Consumers have come to expect that a D2C company is an easy and accessible partner for transactions and support, giving them what they want when they want it.
- **Capital-light ecosystem business models.** One common hallmark of D2C startups is that they require relatively little in terms of conventional capital. They outsource much of the operations, joining ecosystems built on digital platforms, where infrastructure becomes a shared resource. These companies don't compete on better distribution or supply chains—they can put together complex supply chains in a fraction of the time and expense it would take in an analog world. Instead, they compete on what really matters: a better customer experience.

The Theory of Disruption: What Stays the Same

Christensen's original theory of disruption has held up very well in explaining why startups with little in the way of assets or existing brands can capture market share from well-entrenched incumbents. Just as the theory predicted, incumbents considering investments in innovation that has the potential to

These companies compete on what really matters: a better customer experience.

cannibalize the existing business still find it unattractive and dangerous. They have little incentive to pursue opportunities with thinner margins than those enjoyed by their core business, and their corporate metrics tend to reinforce this status quo.

As Christensen also predicted, the "jobs" customers seek to get done in their lives remain remarkably stable[5]—even though digital technologies have created entirely new ways to get those jobs done. Consider the job of making an apartment comfortable by furnishing it. Today's young, nomadic urban workers often find that it's more convenient to accomplish that job by leasing furniture than buying it. Incumbents can get blindsided by this kind of shift in how a job gets done. In particular, they may find that their competitor isn't a traditional one but a company from a different industry altogether that has mastered the new digital technologies. Apple, for instance, is partnering with Goldman Sachs to issue an Apple-branded credit card. You might also think of e-retailer Alibaba's threat to banks with its payment systems, Amazon's move into groceries and brick-and-mortar shops, or Uber's effort to dominate third-party food delivery.

Christensen's theory also holds for the fact that creating new customers by lowering prices enough to compete with nonconsumption is still a viable opportunity for crafty newcomers. Just as traditional disruptive competitors pulled new buyers into new markets by lowering prices, digitally disruptive companies make it radically cheaper, easier, and faster to become customers. Consider, for instance, what has happened in the market for hearing aids. A traditional fitting for a hearing aid required a labor-intensive and extremely expensive visit to an audiologist and a cumbersome process of fitting and adjusting the devices. Eargo, a venture-backed startup, dispenses with all that. Its "invisible" hearing aids (inspired by a fishing fly) fit in your ear. You can fit them yourself. They recharge in a special case—no more hunting down and changing batteries. And the Eargo comes at a lower price point than many traditional hearing aids, potentially opening a vast market of people who need hearing assistance but can't afford the traditional model (especially when hearing aids are not covered by most medical plans).

Finally, Christensen's perspective on what he called the *capitalist's dilemma* is still with us.[6] In many large organizations, incentives are not aligned with the market-creating innovation. One prominent example: massive share buybacks, which handsomely reward executives while draining companies of cash that could be invested in innovation designed to win new customers by transforming exclusive products and services into simpler, inexpensive ones.[7]

The Theory of Disruption: What Has Changed

Christensen described disruption as a process that takes some time, as new entrants slowly progress from the fringe to the mainstream of an incumbent's business.[8] The most significant change since he first laid out his theory is that digital competitors can now move with unprecedented speed.

The conditions for entry into any sector that makes any margin at all have never been better. There's ample available financing (as of this writing, anyway), talent aplenty in the gig economy, consumers who are comfortable buying just about anything sight unseen, and digital technologies to facilitate every operation that might previously have been an obstacle. As Warby Parker, Casper, and the like have shown, disrupters with a competitive value proposition can drive scale at previously unimagined speed.

A second departure from the theory of disruption has to do with the relationship between the traditional, core business and innovative new ones. In the original formulation, the core part of the business had fairly predictable (if slowly declining) revenue numbers, customers whose needs could be identified, and rewards for replicating the existing model at scale. Innovative new businesses, on the other hand, have operated with a high ratio of assumptions relative to knowledge, leading to practices such as discovery-driven planning, test-and-learn, and rapid experimentation.

Today's digital disruption is so fierce that core businesses are less reliable than ever, and their declines can sometimes be precipitous to the point of endangering the entire enterprise. Consider the fate of General Electric, once the darling of admiring business school cases and now described as being "on life support."[9] GE's management realized relatively early on that digital was likely to bring massive change to its businesses. In 2013, it embarked upon a digital transformation with the launch of a platform called Predix, which was supposed to harness the internet of things and bring disruptive change to the storied conglomerate. But GE failed to balance well its investments for the future with the need to meet quarterly numbers. When Predix failed, its demise adversely affected the health of GE's other, core, divisions, leaving the company in dire straits.

There's one more important change that's happened since Christensen's early work was published. Incumbents have learned a thing or two about disruption. Leaders at German metals distributor Kloeckner, for instance, determined that if they didn't create a digital platform for doing business, some upstart would do it to them, and they have been on a steady journey to digitize their industry. Other incumbents are willing to use their resources aggressively to combat disruption. They shell out eye-popping sums to acquire startups, they try their best to import a startup mentality and practices, and they leverage their own resources and heft to let their digital acquisitions or offshoots accelerate to scale. Walmart, for instance, spent $3.3 billion to acquire Jet.com, and millions more to acquire a string of D2C companies whose offerings are appealing to a younger demographic. GM and Ford spent heavily to compete in the emerging sector of autonomous vehicles. Incumbents have read Christensen, and the best ones are doing everything they can to avoid the slothful mistakes of the past.

In general, when seeing a disruption coming, incumbents seem to fall into three categories. The first are those that fall into the classic Christensen trap and ignore the potential change completely. The second are those that spot the disruption and overreact, spending vast amounts of money and time on efforts to jump right into whatever the disruptive market seems to hold. What I would observe is that companies such as Kloeckner that begin to make modest investments in potential disruptions gradually create the capabilities to segue into the next phase without a wrenching downfall or excessive shift. Toyota, for instance, created a mass market for hybrid electric vehicles without abandoning its core internal combustion business, and it is one of the few profitable players in the electric vehicle arena.

The Road Ahead for Incumbent Companies

Every traditional company should be aware that the very concept of sustaining innovation is at risk when a digital assault on the core business is as easy, fast, and affordable as it is today. Digital puts the disruptive mantra of "faster, cheaper, and good enough" on steroids. Business models enabled by digital create potential inflection points for every traditional business. Senior leaders and board members must accept that there are no safe bets.

Yet all too often, business leaders of incumbent companies spend way too much money on digital transformation efforts that fail to take the new economics and business models of digital disrupters into account. Automating old business models is nothing more than that—it doesn't do a thing to help your company benefit from the disruptive price/performance ratios that digital tools can foster.

There is a tremendous amount still to be learned about how to compete in a world moving at the pace of digital. This places a huge premium on being able to learn quickly, experiment, and then pivot to reflect the insights gleaned. Incumbents need to stop spending money trying to be a better version of their analog selves, and instead start approaching digital strategy with an eye toward discovery.

IV

The Futurist Element

12

The 11 Sources of Disruption Every Company Must Monitor

Amy Webb

Recently I advised a large telecommunications company on its long-term strategy for wireless communications. The company was understandably concerned about its future. A half-dozen new streaming TV services were in the process of being launched, and bandwidth-hungry online gaming platforms were quickly attracting scores of new players. Possible regulatory actions seemed to be lurking around the corner, too.

Changes like these meant disruptions to the company's existing business models, which hadn't materially evolved since the dawn of the internet age. As a result, the company worried that it might be facing an existential crisis. To get in front of the risk, its senior leaders wanted to dispatch a cross-functional team to produce a three-year outlook analyzing which disruptive forces would affect the company and to what degree. It was no simple effort. First, the leaders had to galvanize internal support. At this company, any change to standard operations required lots of meetings, presentation decks, and explanations of concrete deliverables. Once they had buy-in and the cross-functional team was in place, they spent months researching the company's competitive set, building financial models, and diving deeper into consumer electronics trends.

Finally, the team delivered on its mandate. A detailed, comprehensive three-year plan projected that new streaming platforms and online gaming would cause a drastic increase in bandwidth consumption, while newer connected gadgets—smartphones, watches, home exercise equipment, security cameras—would see greater market penetration. It was a narrow vision that would take the company down a singular path focused only on streaming and consumer gadgets without considering other disruptive forces on the horizon.

The findings were hardly revelatory. Streaming platforms, gaming, and gadgets were a given. But what about all the other adjacent areas of innovation? In my experience, companies often focus on the familiar threats because they have systems in place to monitor and measure known risks. This adds very little value to long-term planning, and, worse, it can lead to organizations having to make quick decisions under duress. It's rarer for companies to investigate unfamiliar disruptive forces in advance and to incorporate that research into strategy.

I was curious to know how the company had initially framed its project. The objective was to investigate all of the disruptive forces that could affect telecommunications in the future, yet it had really focused only on the usual known threats.

There were plenty of outside developments worth attention. For example, some clever entrepreneurs had already deployed new systems to share the computer processing power sitting dormant in our connected devices. Using a simple app, consumers were selling remote access to their mobile phones in exchange for credits or money that can be spent on exchanges. (This literally allows consumers to earn money while they sleep.) Since the

Companies often focus on the familiar threats because they have systems in place to monitor and measure known risks. This can lead to organizations having to make quick decisions under duress.

systems are distributed and decentralized, private data is safeguarded. On these new platforms, anyone can rent their spare computation resources for a fee.

What's most interesting about distributed computing platforms is that they can also harness the power of other devices, like connected microwaves and washing machines, smart fire alarms, and voice-controlled speakers. As distributed computing platforms move from the fringe to the mainstream, this would have a seismic impact on the telecommunications company's financial projections. While the team was accustomed to calculating the cost per megabit for streaming and the cost to maintain its networks, it didn't have formulas to calculate the financial impact of billions of connected devices that could soon be a part of giant distributed computing platforms.

Looking at the future of telecommunications through the lens of distributed computing, I had a lot of follow-up questions: How should existing bandwidth models and projections be revised to account for all of these devices? Would customer plans still earn the same margins with all these new use cases for existing bandwidth? Would the company mine all of the device data for business intelligence? If so, what would data governance need to look like?

I also asked the team to think about the future of telecommunications through another adjacent lens: climate change. Existing data centers, like all buildings, were developed using guidelines, architectural plans, and building codes that will likely need to change in response to severe weather events. Data centers must be housed inside temperature-controlled environments that never deviate. Heat waves, flash floods, hail, high winds, and wildfires have become more common—and harder to predict. This poses a threat to critical infrastructure.

While the team could build predictive models to anticipate bandwidth spikes, predicting extreme weather events would be far more difficult. How was the team tracking weather and climate? Had they built uncertainty into their financial projections to account for extreme weather events? Was there a crisis plan ready to implement if the power got knocked out? What if a long stretch of exceptionally hot days strained the air conditioners? Did it make sense for the company to continue building and maintaining data centers? Was there a case to be made for adding a small team of climate scientists to the company's existing data science unit?

I could see from everyone's reactions that this line of exponential questioning was beyond the typical scope of their research. The reason the company had not considered these and other areas of potential disruption had to do with its entrenched habits and cherished beliefs. The team was accustomed to a rigorous—but narrow—approach to planning. They built financial projections, tracked their immediate competitors, and followed R&D within their industry sector. That was it.

What I observed is hardly unique. When faced with deep uncertainty, teams often develop a habit of controlling for internal, known variables and fail to track external factors as potential disrupters. Tracking known variables fits into an existing business culture because it's an activity that can be measured quantitatively. This practice lures decision makers into a false sense of security, and it unfortunately results in a narrow framing of the future, making even the most successful organizations vulnerable to disruptive forces that appear to come out of nowhere. Failing to account for change outside those known variables is how even the biggest and most respected companies get disrupted out of the market.

When faced with deep uncertainty, teams often develop a habit of controlling for internal, known variables and fail to track external factors as potential disrupters.

Futurists call these external factors *weak signals*, and they are important indicators of change. Some leadership teams lean into uncertainty by seeking out weak signals. They use a proven framework, are open to alternative visions of the future, and challenge themselves to see their companies and industries through outside perspectives. Companies that do not formalize a process to continually look for weak signals typically find themselves rattled by disruptive forces.

As a quantitative futurist, my job is to investigate the future, and that process is anchored in intentionally confronting uncertainties both internal and external to an organization. I do this using what I call the *future forces theory*, which explains how disruption usually stems from influential sources of macro change. These sources represent external uncertainties—factors that broadly affect business, governing, and society. They can skew positive, neutral, or negative.

I use a simple tool to apply the future forces theory to organizations as they are developing strategic thinking. It lists 11 sources of macro change that are typically outside a leader's control. (See "The 11 Macro Sources of Disruption.") In 15 years of quantitative foresight research, I have discovered that all change is the result of disruption in one or more of these 11 sources. Organizations must pay attention to all 11—and they should look for areas of convergence, inflections, and contradictions. Emerging patterns are especially important because they signal transformation of some kind. Leaders must connect the dots back to their industries and companies and position teams to take incremental actions.

The 11 sources of change might seem onerous at first, but consider the benefit of a broader viewpoint: A big agricultural company tracking infrastructure changes could be a first mover

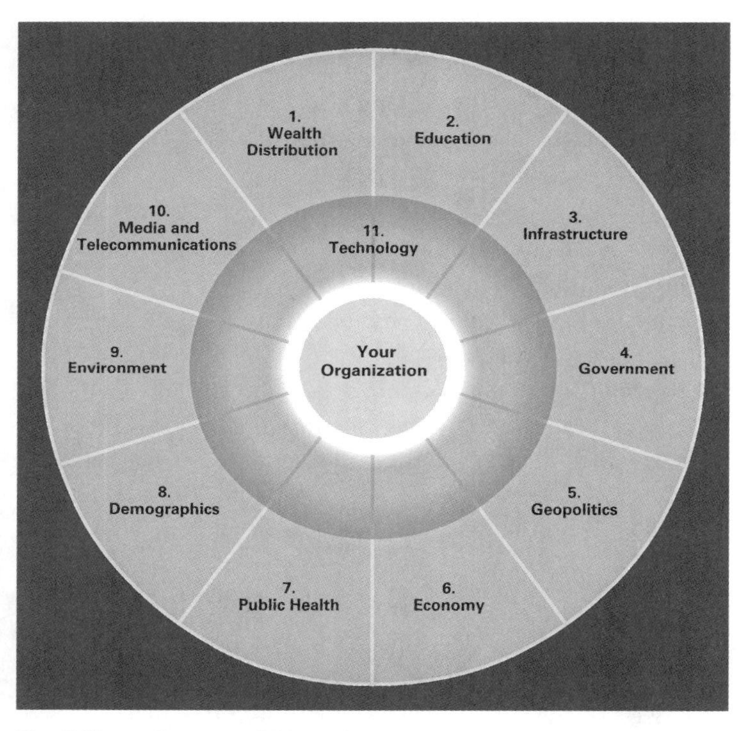

The 11 Macro Sources of Disruption
This simple tool shows the 11 sources of macro change that are typically outside a leader's control. Because technology is so intertwined with everyday life, it is shown as intersecting with all the other sources.

into new or emerging markets, while a big-box retailer monitoring 5G technology and artificial intelligence could be better positioned to compete against the big tech platforms.

Sources of macro change encompass the following:

1. **Wealth distribution:** the distribution of income across a population's households, the concentration of assets in various

communities, the ability for individuals to move up from their existing financial circumstances, and the gap between the top and bottom brackets within an economy

2. **Education:** access to and quality of primary, secondary, and postsecondary education; workforce training; trade apprenticeships; certification programs; the ways in which people are learning and the tools they're using; what people are interested in studying
3. **Infrastructure:** physical, organizational, and digital structures needed for society to operate (bridges, power grids, roads, Wi-Fi towers, closed-circuit security cameras); the ways in which the infrastructure of one city, state, or country might affect another's
4. **Government:** local, state, national, and international governing bodies, their planning cycles, their elections, and the regulatory decisions they make
5. **Geopolitics:** the relationships between the leaders, militaries, and governments of different countries; the risk faced by investors, companies, and elected leaders in response to regulatory, economic, or military actions
6. **Economy:** shifts in standard macroeconomic and microeconomic factors
7. **Public health:** changes occurring in the health and behavior of a community's population in response to lifestyles, popular culture, disease, government regulation, warfare or conflict, and religious beliefs
8. **Demographics:** observing how birth and death rates, income, population density, human migration, disease, and other dynamics are leading to shifts in communities

9. **Environment:** changes to the natural world or specific geographic areas, including extreme weather events, climate fluctuations, rising sea levels, drought, high or low temperatures, and more. Agricultural production is included in this category.
10. **Media and telecommunications:** all of the ways in which we send and receive information and learn about the world, including social networks, news organizations, digital platforms, video streaming services, gaming and e-sports systems, 5G, and the boundless other ways in which we connect with each other
11. **Technology:** not as an isolated source of macro change, but as the connective tissue linking business, government, and society. We always look for emerging tech developments as well as tech signals within the other sources of change.

This may seem an unreasonably broad list of signals to track to prepare for the future, but in my experience, ignoring these potential sources of change leaves organizations vulnerable to disruption. My favorite example of what comes to pass when companies ignore these signals happened in 2004, when there were a number of emerging weak signals that pointed to a drastic shift in how people communicated. Two senior leadership teams had access to the same information. One looked for external factors actively, while the other simply used trends within its industry to make incremental improvements to its existing suite of products. Those decisions would result in the end of one of the world's most loved and respected companies and the rise of an unlikely competitor that no one saw coming. The signals included the following developments:

- New software made it easy for anyone to rip content from CDs and DVDs.
- Peer-to-peer file-sharing websites like BitTorrent, isoHunt, The Pirate Bay, and LimeWire that were first used by hackers had become popular with ordinary people who were sharing music and movies widely.
- Demand for digital content was growing fast; sales of physical media were starting to decline.
- Game developers were experimenting with haptic technology that responded to pressure and touch. In a combat game, for instance, when players got hit by enemy fire, they'd feel the controller buzz. Developers were also building haptics into early touch screens: Players could simply touch an icon to advance, move back, turn, or stop.
- In Korea and Japan, consumer gadgets were being built with dual functions: There were digital cameras with MP3 players; cell phones had retractable metal antennas to receive broadcast TV signals.

One of the senior leadership teams connected those signals with its existing work and foresaw a world in which all of our existing devices converged into just one mobile phone that had enough power to record videos, play games, check email, manage calendars, show interactive maps with directions, and much more. That team had no cherished beliefs about the existing form factor of our mobile phones and was willing to accept alternative ideas for how a computer-phone could work. That team worked at Apple, and in 2007, a product that had baked all of those weak signals into its strategy went on sale: the first iPhone. By the end of the decade, a company that once was mostly known for its sleek desktop computers had forced the entire mobile device market to bend to its vision of the future.

By contrast, these very same weak signals never caught the attention of Research in Motion (RIM), which at the time made the world's most popular phone, the BlackBerry. (In fact, we loved their phones so much we called them *crackberries* and were proud of our digital addictions.) It was the first device that allowed us to stay truly connected to the office. Perhaps most important, it had a full, physical keyboard. All other phones at that point simply had numbered buttons; to type letters required hitting a few buttons to access one of the three letters assigned to each number. Before the BlackBerry, a simple three-line text message could take several minutes to type.

Because of the BlackBerry's enormous popularity, RIM had become one of the largest and most valuable companies in the world, valued at $26 billion. It controlled an estimated 70% of the mobile market share and counted 7 million BlackBerry users. With its great run of success, the organization's culture did not allow for alternative versions of the future, and internally, there was an aversion to contradicting cherished beliefs. Managers who did connect those weak signals to the BlackBerry didn't have credibility outside their departments. As a result, all of the disruptive external forces Apple was actively tracking never broke through to the senior leadership team of RIM. RIM continued innovating narrowly, selling a smaller BlackBerry Pearl with a tiny, pearl-shaped mouse embedded in the keyboard and releasing BlackBerries in new colors. It was, in hindsight, the defensive strategy that Clayton M. Christensen explained in his Theory of Disruptive Innovation. Threatened by a disruption, incumbents retreat to the strategy of what Christensen called *sustaining innovations*—new bells and whistles that allow the incumbent to keep its customer base and, more importantly, its profit margin. But such innovations virtually ignore the disruptions breaking into the incumbent's market.

Once the iPhone launched, Apple kept listening for signals while RIM never recalibrated its strategy. Rather than quickly adapting its beloved product for a new generation of mobile users, RIM continued tweaking and incrementally improving its existing BlackBerries and its operating system. That first iPhone was in many ways a red herring. As is so often true with successful disrupters, the first product to break through is often low quality and barely "good enough" for consumers. That's what enables incumbents to justify ignoring them. But the ascent to quality is rapid. Apple swiftly made improvements to the phone and the operating system. Soon it became clear that the iPhone was never intended to compete against the BlackBerry. Apple had an entirely different vision for the future of smartphones—it saw the trend in single devices for all of life, not just business—and it would leapfrog RIM as a result.

The ways in which RIM and Apple planned their futures are what sealed their fates, and what happened to RIM is a warning that applies to every organization. Senior leaders can choose to lean into uncertainty and methodically track disruptive forces early, or they can choose to innovate narrowly and reinforce established practices and beliefs.

Many companies around the world use the future forces theory to help them make sense of deep uncertainty and break free from the tyranny of narrow innovation. Some use it at the start of a strategic project, while others use it as a guiding principle throughout their work streams, processes, and planning. The key is to make a connection between each source of change and the company and also to ask questions like: Who is funding new developments and experimentation in this source of change? Which populations will be directly or indirectly affected by shifts in this area? Could any changes in this source lead to future regulatory actions? How might a shift in this area lead to shifts in other sectors? Who would benefit if an advancement in this source of change winds up causing harm?

I have seen the most success in teams who use the macro change tool not just for a specific deliverable but to encourage ongoing signal scanning. One multinational company took the idea to a wonderful extreme: It built cross-functional cohorts made up of senior leaders and managers from every part of the organization all around the world. Each cohort has 10 people, and each person is assigned one of the sources of macro change, along with a few more specific technology topics and topics related to their individual jobs. Cohort members are responsible for keeping up on their assigned coverage areas. A few times a month, each cohort has a 60-minute strategic conversation to share knowledge and talk about the implications of the weak signals they're uncovering. Not only is this a great way to develop and build internal muscles for signal tracking, it has fostered better communication throughout the entire organization.

It might go against the established culture of your organization, but embracing uncertainty is the best way to confront external forces outside your control. Seeking out weak signals by intentionally looking through the lenses of macro change is the best possible way to make sure your organization stays ahead of the next wave of disruption. Better yet, it's how your team could find itself on the edge of that wave, leading your entire industry into the future.

13

The Uncertainty Factor

Rahul Kapoor and Thomas Klueter

For the past two decades, companies have assumed that they know the disruption playbook. It's an S curve of progress: a series of cumulative advances as a new value proposition progresses to outperform a given industry's prevalent offers. A company introduces gradual improvements in a new, innovative value proposition. Initially, the offering is not attractive to mainstream users and established incumbents, but eventually it becomes good enough and then achieves market dominance. Disruption of the incumbent is complete.

This perspective on disruption provides a valuable guide with respect to how investment returns on innovative efforts may unfold over time. Progress during early efforts tends to be slow, followed by takeoff and a period of sustained growth.

The launch of Netflix's DVD-by-mail service at the turn of the century represents a classic example. The service was initially targeted at movie enthusiasts who were early DVD adopters. These were consumers who agreed that the trade-off of selecting films through online search was worth the wait (often several days) for the movies to arrive in red envelopes in the mail. At the time, this value proposition was not attractive compared with the mainstream video rental market. However, as Netflix improved its offer—via an unlimited subscription service, an online recommendation engine, a more efficient distribution network, and newer and original content—the company was able to disrupt video rental incumbents such as Blockbuster.

But this view of disruption is oversimplified or, at a minimum, incomplete. What the prevailing imagery for S-curve progress misses is the fact that there is *significant uncertainty* regarding the rate of progress within the new disruptive value proposition. Some innovations can reach mainstream status in a matter of years, whereas others may take decades. And others, despite their disruptive potential, may never reach fruition. Video streaming services took off rapidly around the globe in a matter of years. In contrast, it has taken online degree programs more than a decade to establish a strong foothold in the education sector. And gene therapy, touted as a major advance in personalized medicine for several decades, has yet to take off.

The importance of factoring in uncertainty to understand the trajectory and impact of a disruptive value proposition on either a startup or an incumbent can't be overemphasized. When it's not anticipated, an otherwise promising upstart might leap forward with a product or service based on the assumption of a strong product-market fit without ever finding its audience. Or an incumbent can find itself taking its eye off its bread-and-butter existing products and services based on the assumption that a new innovation will skyrocket to success and provide the growth engine for the future. Neither path is a good one.

Uncertainty and disruption are two sides of the same coin; they can't be separated. And yet much of the established thinking around managing disruption focuses on incumbents grappling with the threat of market incursions and identifying opportunities to develop their own, and on new entrants managing the opportunities around disruption.[1]

Although it's true that the progression of each disruptive innovation may be shaped by the specific strategies of incumbents and entrants, decision makers should bear in mind that there is substantial uncertainty around whether a disruptive value proposition will materialize in the first place. Why is this important? Because an explicit consideration of uncertainty can help decision makers recognize the risks that surround the execution of the disruptive strategy. It can help them set more realistic market-growth expectations and evaluate strategic contingencies that can be experimented with and validated.

The Analysis

- The authors have analyzed the progress, successes, and failures of disruptive innovation efforts in sectors such as health care and energy.
- They have also surveyed the academic literature and publicly available reports on disruptive innovation in other industries.

In our ongoing research, we have found three key sources of uncertainty—around technology, ecosystems, and business models—that are pivotal to understanding the process of disruption. When companies carefully consider these sources of uncertainty and how to address them, they can better position themselves to manage disruption and achieve superior performance outcomes. When entrepreneurs and executive teams overlook these factors, it opens up their companies to foreseeable challenges, such as the following:

- Failing to recognize the time and the extent of resources that might be required for the disruptive value proposition to take hold. This can lead to misjudgments about investments in a disruptive innovation initiative, such as giving up too early or sustaining significant spending too long, or starving other, more viable, initiatives of resources and attention.

- Focusing on the new technology or the new business model while overlooking the challenges within the company's ecosystem of suppliers, business partners, and customers that may be critical to the realization of the new value proposition. This can lead to prematurely optimistic projections about the potential of a disruptive innovation and risks wasting resources.
- Missing opportunities they could otherwise identify and seize around business model innovation across different markets. The risk here is that a company limits the potential appeal of a disruptive innovation or narrows the innovation's paths to market without examining all the possible variations around the business model.

These uncertainties do not influence every company to the same degree, of course. Startups tend to be adept at experimenting with new technologies and business models, even though they may be resource-constrained. In contrast, established companies tend to be endowed with significant resources but face significant adjustment costs when they pursue disruptive value propositions while managing their core business. (Although startups and established companies often compete, collaboration can help both manage the uncertainties of disruption. See "The Potential for Collaboration in the Face of Uncertainty.") But without a deep understanding of how uncertainty can affect the speed and resource-intensiveness of the disruptive arc of development, startups and incumbents alike can find themselves failing at what otherwise might have been a successful disruptive innovation.

The Potential for Collaboration in the Face of Uncertainty

Disruptive innovation typically is framed as a contest in which the startups threaten and, potentially, displace the established industry players. Considering the different sources of uncertainty and the unique challenges each type of company confronts, our research suggests that collaboration between startups and incumbents can help advance disruptive innovations in areas such as gene therapy and decentralized electricity, where the innovation has the potential to be truly disruptive but needs time and resources to find its value proposition, business model, and supportive ecosystem. In these cases, we see a win for incumbents and startups alike. Indeed, they can work together to address the constraints and overcome the challenges that are specific to each actor and, at the same time, present complementary opportunities for actors to pursue joint value creation. This can have at least three benefits:

- Established companies and startups can pool resources and share risks during a period of significant uncertainty.
- Startups can help established companies experiment and validate new business models.
- Established companies can leverage their ecosystems and resources to help startups scale up their disruptive innovations.

Such patterns of collaboration are now rampant in the automotive sector, with emerging entrants such as Aurora, Grab, Uber, and Waymo and established automotive players such as Daimler, Ford, General Motors, and Toyota jointly pursuing autonomous vehicles.

Three Sources of Uncertainty

How is it that otherwise savvy companies tend to overlook or ignore potential sources of uncertainty? They might not be looking in the right places, or they may be locked into a specific strategic perspective too early. Our research has identified three key sources of uncertainty surrounding the question of whether the disruptive value proposition will reach fruition in a given market.

1. **The enabling technology.** Questions can persist about whether the technology that is enabling the disruptive value proposition can achieve the performance-cost threshold required for adoption by mainstream users (that is, for achieving product-market fit). For example, for commercial space travel, there are technological questions related to performance and cost, and although a large number of companies are pursuing the new value proposition, it remains unclear which technological design may be most feasible.
2. **The surrounding ecosystem.** Uncertainty may also stem from not knowing whether actors in the ecosystem will contribute to the disruptive value proposition through supporting investments, complementary innovations, or standards and regulation. For example, there remain important gaps in understanding regarding the use of augmented reality for instruction and training; it's unclear whether there will be sufficient complementary content and suitable hardware devices for users to benefit from the new value proposition and how such virtual offerings might be regulated.
3. **The business model design.** Finally, there can be unsettled issues around the viability of the business model. Will the revenue and profit streams reach sustainable levels for the companies pursuing the disruptive value proposition? For example, there is significant uncertainty around whether the business model for autonomous vehicles looks more like

traditional private and fleet vehicle ownership or like a fee-based mobility-as-a-service offering.

These uncertainties are not isolated. As our research on gene therapy has revealed, they can sometimes combine to heighten the challenge of commercializing innovations with disruptive value propositions.[2]

Gene Therapy's Suspenseful Story

Gene therapy has faced challenges since its emergence in the 1980s. It has the potential to be a game changer for patients with genetic disorders that have no known cures, because it promises to cure the diseases by fixing defective genes instead of treating symptoms. When seen through the lens of the three uncertainties, however, it's clear that the path to disruption is a steep climb.

Early attempts in gene therapy development proved ineffective, and some clinical trials led to severe patient side effects and deaths, raising questions about the time and resources required to bring this innovation to market. The business model for gene therapy is also unsettled. Because the treatments can mean a permanent cure or less-frequent treatments than prevailing methods, calculating pricing and insurance reimbursements has proved difficult. Companies have discussed several business models, including a pay-per-treatment approach, payments spread over a fixed time line, or a pay-for-performance model in which payments are halted if the treatments have stopped working.[3] Gene therapy also confronts significant ecosystem uncertainty. Treatments need to be administered by trained physicians in specialized settings, and they need to be reimbursed by insurance plans. But the availability of trained physicians, gene therapy facilities, and insurance plans that provide coverage is difficult to establish.

Even when a gene therapy treatment wins approval, it can face a cloudy future. A recent case illustrates the challenge: Gene therapy company uniQure pursued a treatment for lipoprotein lipase deficiency, a rare disorder that prevents a person who lacks certain proteins from breaking down fat molecules. The company and health care insurers found it difficult to price the treatment for such a small patient population. UniQure, which had won European approval for the treatment, subsequently withdrew it because of the pricing challenges and the rarity of the disease.[4] The fate of this treatment is emblematic of the obstacles other gene therapy companies face. The disruptive potential of their innovations is enormous, but even after a company overcomes the technological and R&D hurdles and creates a new offering, it must confront business model uncertainties.

The life-and-death implications of gene therapy as a potential disruption make it a dramatic example. But this analysis of uncertainties is applicable whenever products and services with a disruptive value proposition emerge. Using this lens to assess the circumstances in which their company enters a particular market enables leaders to make more precise decisions about resource allocation and timing—and guides their expectations about returns.

Considerations for Established Companies and Startups

How might established enterprises and startups be affected by these sources of uncertainty? Their motivations for pursuing disruptive innovations differ, and their strategies are shaped by

While nearly all disruptive startups are motivated by the possibility of replacing the industry's status quo, many of them confront resource constraints in their efforts to develop a disruptive value proposition.

their available resources and how they measure performance. So each type of company must weigh different considerations.

While nearly all disruptive startups are motivated by the possibility of replacing the industry's status quo, many of them confront resource constraints in their efforts to develop a disruptive value proposition. In the case of gene therapy, startups have attracted a lot of attention, but many could not continue in the face of technology setbacks as their resources and new sources of funding dried up. Highly promising gene therapy startups like Introgen Therapeutics and NeuroLogix ultimately filed for bankruptcy in the United States.

Established market leaders face other challenges. Although they typically have significant resources available to explore disruptive innovations, they cannot focus solely on this quest. They have to simultaneously manage their core business and measure progress against prevailing key performance indicators (KPIs) and short-term investor expectations. And established companies also may be industry incumbents facing a direct threat from a disruptive innovation or from players active in adjacent industries who see their own opportunity to grow in a related industry at the incumbent's expense. In our research, we saw evidence of several established pharmaceutical companies, such as GlaxoSmithKline and Merck, investing in gene therapy research but holding back its commercialization because of business model and ecosystem uncertainty.

Analyzing a company's resource availability and the need to manage performance carries over to the three key uncertainties for disruption.

Resolving technological uncertainty requires significant resources over time to achieve the performance-cost threshold necessary for product-market fit. Given that startups tend to be

Failure to account for critical actors such as regulators and creators of complementary innovations can cause progress bottlenecks and constrain the value proposition of the disruptive innovation.

resource-constrained, they may be more adversely affected by this type of uncertainty. For example, in the case of companies pursuing new solar power technologies, many promising startups had to exit the industry once they lost the technology race to alternative solutions, whereas many established companies were able to continue directing significant resources toward the emerging market opportunities.[5] For example, Solyndra entered the renewable energy market with a promising solar power technology called copper indium gallium selenide, but it ended up losing the battle for market dominance to crystalline silicon, resulting in an abrupt bankruptcy.[6]

Resolving ecosystem uncertainty represents a coordination dilemma, given that business leaders need to manage significant investments across multiple actors, including business partners, suppliers, customers, and regulators. Failure to account for critical actors such as regulators and creators of complementary innovations can cause progress bottlenecks and constrain the value proposition of the disruptive innovation. In such situations, startups may have a steeper challenge: Not only are they resource-constrained, but they may also lack scale and credibility among members of the ecosystem to influence their supportive actions.

Consider the case of Better Place, with its disruptive value proposition around electric cars. Its model to offer battery-charging and -swapping services in addition to selling vehicles helped resolve the technological uncertainty for motorists for whom the low battery performance and high cost of electric cars did not offer a strong value proposition. But as it pursued its growth trajectory across different geographies, Better Place was unable to orchestrate the ecosystem and align the different actors—including customers and the governments in its targeted markets of Denmark and Israel—and sold only about 1,300 cars before going bankrupt in 2013.[7]

Autonomous vehicles are another example of a potential auto industry disrupter, with a number of established automakers launching deliberate, collaborative efforts to develop an ecosystem. BMW, for one, is working with Mobileye (Intel's vision-safety venture) and Fiat Chrysler, as well as parts suppliers like Aptiv, Continental AG, and Magna International, with the goal of commercialization by 2021.

Resolving business model uncertainty requires continuous experimentation and the ability to reconfigure one's approach to unlock the potential of the disruptive innovation for the innovation's users and the innovating companies. Established companies are more likely to struggle with such uncertainty because experimenting with new profit formulas runs counter to existing metrics. Executives face pressure to meet KPIs. Scrutiny by investors and analysts, meanwhile, typically rewards sustaining rather than disrupting profit models. Conversely, startups are not entrenched in prevailing business models and may be better equipped to manage business model uncertainty.

The recent struggles of electric utilities to adapt to more decentralized business models exemplify the challenges for established companies. In a decentralized model, users (such as homeowners) consume electricity that is generated at or near the point of use, often through a combination of rooftop solar photovoltaic systems, batteries, and digital management of the electricity grid. Incumbent companies that were entrenched in the old, centralized business model had lower performance outcomes when pursuing the disruptive value propositions.[8]

For example, NRG, a US-based energy incumbent, reported large losses from its pursuit of a decentralized model, resulting in the CEO's firing. His departure was followed by a number of articles in the trade press describing the internal conflicts between the centralized and decentralized businesses, unforeseen delays, cost overruns during the implementation of the decentralized model, and the extensive competition NRG faced from new entrants.[9] Incumbent energy companies elsewhere, such as AGL in Australia and RWE in Germany, have faced similar challenges.

Managing the Uncertainty of Disruption

Pursuing a disruptive innovation means taking on risk that the effort may fail. Analyzing the uncertainties that any disruption faces, however, can help you mitigate those risks by making informed decisions about the supporting technology, the surrounding ecosystem, and the business model foundation required for success.

The following five questions can help innovators—incumbent companies and startups—manage uncertainty.

1. **What are the opportunities for a disruptive value proposition?** Opportunities can be related to creating new markets, such as space tourism, or penetrating existing markets, such as global tourism.
2. **Where are the key sources of uncertainty – technology, ecosystem, and business model – in different markets?** Uncertainty doesn't have to be prevalent across all areas. It is typically a subset of the three sources that can create bottlenecks for market growth.
3. **How can the different sources of uncertainty be addressed?** Experimentation with respect to customers and others in the ecosystem, business models, and technology choices can be

Based on the answers to five questions, leaders can make more nuanced decisions about their plans, including which innovations to pursue, how much to invest, which partners to collaborate with, and the timing for all of these choices.

valuable in resolving uncertainty. However, if uncertainty is severe across the three sources, decision makers may need to say no to investments at the outset or stop specific disruptive innovation initiatives.

4. **Can I pursue this disruption on my own, or do I need strategic partners in the ecosystem to help resolve uncertainty?** Identification of ecosystem activities and actors where uncertainty resides—and coordination among them—can be a critical aspect of managing such uncertainty.
5. **How can I align partners to cocreate value?** Partners can have different business models and motivations around the disruptive value proposition. It is important to identify which partners may have mutually beneficial objectives and to ensure that those objectives are aligned for the long run.

This exercise widens leaders' perspectives on the opportunities before them. By openly considering these questions, leaders improve their ability to identify the risks of any strategy to develop a disruptive innovation. Based on the answers to these questions, they can make more nuanced decisions about their plans—which innovations to pursue, how much to invest, which partners to collaborate with, and the timing for all of these choices—than they otherwise would. They can revisit their evaluations and adjust their investments and the timing of them. And they can calibrate their expectations for progress on a particular value proposition and whether it has the potential to disrupt an established market based on additional evidence and insights.

Well-established cases show what's possible for both established companies and startups. For the iPhone, Apple managed ecosystem uncertainty (What parties will work with us on this

To prepare for the best possible outcome, it's important to understand not only the reach of an innovative idea but also the risks that lie in its path to realization.

disruption? How?) through the creation and maintenance of its App Store and its calibrated rollout of available telecommunications carriers. Apple also navigated business model uncertainty in part through its use of exclusive vendors when rolling out the iPhone. For its still-evolving electric car business model, startup Tesla has managed technology uncertainty by investing in batteries and software in order to offer a high-performance electric car. Regarding questions about its business model, Tesla set up direct sales. And the company is working to manage ecosystem uncertainty (how to keep electric cars charged) by developing infrastructure through initiatives like its Supercharger network of charging stations.

Disruptive innovations have made us more productive, better informed, and more mobile. They improve our health. They entertain us. And there are many more disruptive innovations to come—indeed, the next one may be at your company. To prepare for the best possible outcome, it's important to understand not only the reach of an innovative idea but also the risks that lie in its path to realization. With our eyes open to confront uncertainties around the technology, ecosystem, and business model of each potential disruption, we can better understand what to expect along the way and better devote the time and resources to strengthen our chances of success.

14

Why Innovation's Future Isn't (Just) Open

Neil C. Thompson, Didier Bonnet, and Yun Ye

New digital technologies have upended conventional business models, organizational structures, and operating processes in most industries. Almost every aspect of business—customer relations, supply chain management, after-sales service—has been radically altered.

Nowhere is that more evident than in brick-and-mortar companies' innovation processes. Facing tough competition from digital upstarts that are creating and capturing value in new ways, incumbents are trying to figure out how to keep up.

While the need for innovation in the digital age may be an open-and-shut case, CEOs of these companies aren't sure whether their innovation processes should be open, shut, or both. Many businesses that used to depend only on internal innovation have begun to tap external innovation to quickly acquire the digital capabilities they need to navigate the constant stream of new technologies. However, forging partnerships and investing in startups doesn't always confer a competitive advantage; often, we find, it only helps incumbents catch up with rivals. The need to acquire capabilities quickly is important, but companies must also continue to invest in developing key capabilities internally, even if that takes more time.

Most companies, our research suggests, should rethink their innovation systems and develop portfolios with a balance of innovation sources. They must treat external innovation as a way of broadening their portfolios, not as a substitute for internal innovation. Only then will they be able to execute the transformations that will allow them to win the digital future.

The Research

- The research behind this article, conducted in 2018–2019, included companies in seven industries from Australia, China, France, Germany, Japan, South Korea, the United Kingdom, and the United States.
- The authors held in-depth interviews with executives in 30 large corporations across industries and countries to obtain a granular understanding of innovation practices and systems.
- They then structured and, through Phronesis Partners, administered a survey to quantify innovation practices and systems, polling innovation leaders at 320 large corporations with revenues of more than $500 million a year and gathering data on 640 innovation projects.

Reaching Out to New Partners

It's beginning to dawn on many businesses, be they consumer-facing or industrial, that digital transformation isn't just about investing more in IT, digitizing operations, and creating websites, mobile apps, and online channels to connect with customers and suppliers. All that improves efficiency, but it doesn't alter the business model. A digital transformation is a deeper change. It entails using second-wave technologies such as the industrial internet of things, artificial intelligence, and machine learning to create new sources of value through product and process innovation. For instance, manufacturers are figuring out how to use robotics and machine learning in production and supply chain management, and services companies are learning to deploy augmented and virtual reality to improve training.

But transforming a business is hard, expensive, and time-consuming, so companies are dedicating more and more resources to it. In 2012, experts recommended creating an innovation portfolio that allocated 10% of the budget to transformational projects, 20% to moderately transformational projects (sometimes called adjacent or substantial projects), and the remaining 70% to incremental innovation. By 2018, our studies show, most companies were focusing squarely on transformation, investing as much as 30% of their outlays in major projects and 35% in moderately transformational projects. Only 35% of budgets were being spent on incremental innovation—half the percentage in 2010.

What's more, 95% of the companies we surveyed said their most successful innovation project in the recent past had been mainly digital. That number was remarkable not only because it was so high but also because it was so similar across industries.

Yet many of our respondents also admitted that they didn't possess the capability to develop digital applications. Across all types of innovation projects, companies reported that their internal capabilities weren't as good as those of the market leaders 51% of the time. But when it came to digital innovation, they fell short of the leaders an overwhelming 81% of the time. That's why so many companies are looking to source innovation from outside the organization; they are desperate to access digital capabilities and applications—quickly.

Although companies have been experimenting with open innovation for more than 15 years, our data reveals a notable shift in recent years. Companies have begun to seek innovation from a wide variety of potential partners, not just the ones with which they've already built relationships. Just seven years ago, 70% of businesses worked only with the existing partners in their value chains, either customers or vendors. But in the past five years, more than 75% of the companies we surveyed started drawing on a multiplicity of innovation sources. Nowadays, only 17% limit themselves to supply chain partners; most work with several kinds of partners, such as universities, think tanks, consultants, crowdsourcing platforms, startups, and innovation labs (which are internal structures that serve as bridges to external innovation).

Businesses in a wide variety of traditional brick-and-mortar industries, such as manufacturing, retail, and even fast food, are tapping diverse sources of digital innovation. McDonald's, the world's second largest fast-food chain, is an illustration of this shift. In the past two years, it has invested in developing a mobile app and a mobile-based order system as well as self-order kiosks inside its restaurants using innovations sourced from outside the company.

After investing in a mobile-app vendor, Plexure, in early 2019, the company started experimenting with digital personalization at drive-through locations. In April 2019, McDonald's acquired Israeli startup Dynamic Yield, whose personalization systems and decision-logic technologies allow the company to personalize its outdoor drive-through menu displays based on the time of day, the weather, current restaurant traffic, and trending menu items. The technology can also display suggestions from the menu based on a customer's selections. Dynamic Yield's technologies were integrated into nearly all McDonald's drive-through windows in the United States and Australia by the end of 2019.

McDonald's has also focused on digital voice recognition. In September 2019, it acquired Apprente, a startup developing voice-based technology that can handle multilingual, multiaccent, and multi-item ordering. That will allow for faster, simpler, and more accurate order taking at drive-throughs, and, over time, the technology will be incorporated into mobile ordering and the kiosks inside McDonald's outlets. In fact, the Apprente team will be the first members of a new group, McD Tech Labs, in McDonald's Global Technology team, whose goal is to grow the company's footprint in Silicon Valley.

The Benefits of External Innovation

Becoming more open to external innovation is a marked departure from the "not invented here" syndrome from which companies suffered for decades. That old, centralized approach to innovation has its limits; as Sun Microsystems cofounder Bill Joy famously said, "No matter who you are, most of the smartest people work for someone else." Companies now recognize that when they lack internal capabilities that will allow them to remain competitive in the digital age, it's essential to look outside the organization for new technologies and applications.

Many companies also look outside to scan for the "unknown unknowns" that could disrupt their businesses in ways they haven't foreseen. Most are trying to make up for lost time, and investing in partnerships is a faster way to tap new technologies than starting from scratch.

Acquiring and developing capabilities externally helps companies mix and match from several sources to meet their

Becoming more open to external innovation is a marked departure from the "not invented here" syndrome from which companies suffered for decades. That old, centralized approach to innovation has its limits.

needs. For example, iFlytek, a global leader in computer speech technology, has set up joint laboratories with several Chinese universities. The director of each laboratory is a university professor, and iFlytek experts are sent to work with them to develop new technologies and applications. This innovation ecosystem allows the company to tap cutting-edge knowledge from diverse sources even as it works on the innovations it needs.

Indeed, using external partners often makes it easier for companies to adopt a cautious approach to innovation. They can experiment with the use of high-potential technologies outside the organization until they are de-risked enough to be incorporated into the company's products and services.

Finally, and importantly, open innovation helps guard against one of internal innovation's biggest risks: incrementalism. Innovators inside an organization, particularly in the business units, tend to develop products and processes that are easiest for the organization to adopt. That effect can be so strong that it stymies the creation of radical innovations that will reinvent the existing business. Incrementalism will prevail, pushing change forward in small steps that will leave the company vulnerable to disruptive threats from outside.

The Power of Doing It Yourself

Given those factors, the future of innovation may appear, inevitably, to be open. But our research suggests that open innovation need not, and should not, be the only way forward.

Indeed, internal innovation may be even more critical because it offers the possibility of differentiation. Leading digital companies—whether they're technology vendors, application developers, or data giants—like to work with many brick-and mortar companies on key applications in order to spread development costs, attract venture capital, and become industrywide platforms. For instance, Google's Waymo, which is developing autonomous driving technology, has struck partnerships with Fiat Chrysler and Jaguar Land Rover as well as Renault-Nissan. That will give all three global automakers the benefit of offering consumers self-driving cars, but none of them will enjoy a major edge over the others.

That's the key challenge of depending too heavily on external innovation: Because such technology partnerships don't deliver differentiation—they tend to simply raise the baseline and establish parity between rivals—they confer less competitive advantage than internal innovation. Our studies bear this out. Of the innovation projects that our sample of companies conducted internally, 87% yielded persistent or sustained competitive advantage, whereas only 60% of the innovation projects sourced externally did so. Put another way, only 13% of in-house innovation yielded no advantage or an advantage that rivals quickly matched, but that happened 40% of the time when companies depended on external innovation sources.

Internal innovation is critical for digital transformation for many reasons. Apart from delivering competitive advantage, it helps maintain trade secrets and protect intellectual property. Doing that becomes much more difficult when working with partners; that's true even if your partners don't work with your direct competitors, because they may still percolate what they've learned through the industry. Even if a technology or knowledge isn't shared with rivals, problems can still occur. When a company sources a key capability externally, that supplier becomes central to its value-creation process. For example, if a machine-learning startup figures out what product- or feature-usage data best predicts renewals and upselling opportunities, it can wrest bargaining power—and a large share of the profit pool—from its partners.

Further, internal innovation allows experiments to be done in real-life situations. Born-digital companies, such as Facebook and Booking.com, are constantly testing different algorithms and designs. In doing so, they can gather real consumer data, rather than hypothesized responses, in ways that no external provider could offer.

Because internal innovators are familiar with their company's operations, their innovations are often easier to develop, produce, and sell than externally sourced innovations are. Project by project, internal innovations also tend to be more successful and more essential to the business.

Perhaps above all, developing innovation capabilities in-house improves them *durably*, which will stand companies in good stead in the future. In contrast, open innovation will never alter a company's capabilities or culture enough to bring about a transformation.

For all these reasons, a balance must be struck.

Developing capabilities quickly through external sources is critical, but so is tapping knowledge about your customers, your processes, and your culture in a way that only an internal innovation team can do. Smart companies, we find, treat external sources of innovation not as a replacement for internal sources but as a way of broadening the portfolio. They are not mutually exclusive, but complementary. In our survey, the companies that used the most external sources also used the most internal ones. Even companies with several innovation partners reported that their internal sources were the most critical.

Balancing Internal and External Innovation

Since open innovation helps companies acquire capabilities in the short run, but building capabilities internally is the best way to gain competitive advantage in the long run, how do you achieve the right balance? We suggest a three-step approach:

Step 1: Identify critical competencies. To start, companies must determine which technological capabilities are likely to be critical in the future. Some of that happens during the regular strategy-planning process. Most businesses conduct an annual gap analysis of the capabilities they lack, and there are board- and business-level discussions about whether they should be plugged. Rarely does this exercise result in a road map showing which capabilities should be developed internally in the medium or long term and which must be acquired immediately through external sources. That's the missing link.

Of course, a key element in the calculation will be whether a given capability will help differentiate the company from rivals. The degree to which digital technologies are critical will differ; accessing data science expertise may be essential for a chemicals manufacturer, for instance, but it might not be for a real estate management company that needs to understand only sales and rental trends.

Companies must also figure out which external sources will allow them to access the critical competences and applications they've identified. They should reach out to universities, startups, and others to see who is conducting the most exciting innovation relevant to them and then build a portfolio to fill their competency gaps. Keeping abreast of numerous would-be sources of external innovation can be difficult, requiring focused attention and dedicated time by seasoned executives.

Step 2: Create an architecture. Companies must also rebuild their innovation architecture so that they can manage both internal and external sources. It's important to get three building blocks right.

First, most companies will have to refine their organization design. For instance, if startups will be an external innovation source (they usually are), the business must create a way to manage those relationships—such as an incubator, an innovation sandbox, or a venture fund—and its investments in them.

Second, the innovation process must change if the company's powerful business units are to buy into and adopt external innovations. One catalytic structure is an innovation lab in which a company can colocate researchers to gain access to the capabilities of innovation ecosystems in places such as Silicon Valley or Shenzhen. Such labs can be staffed by employees seconded from the company's businesses, which helps get buy-in for external innovations. Interestingly, between 2014 and 2017, the number of corporate innovation labs being built nearly doubled, reflecting the interest in this model.

Finally, companies must develop innovation governance models with appropriate metrics to ensure consistency with their strategy. Many of the companies we studied initially struggled with governance and metrics. They assigned people to innovation projects, but the business units controlled the budgets and approvals. That resulted in slowing down the digital innovation unit, which was hamstrung by the bureaucracy.

Step 3: Develop transfer processes. One of the most common mistakes companies make is not laying down a competence-transfer strategy from the very outset. They must map out how externally developed capabilities and skills will eventually be

Percent of Respondents Who Said These Sources of Innovation Were Important

	At the company level	At the project level
Central R&D	56%	34%
Innovation Lab	13%	25%
Business Unit Staff (dedicated full time to innovation)	17%	15%
Business Unit Staff (operational)	2%	4%
Universities	2%	9%
Crowd	0%	1%
Suppliers	3%	7%
Customers	3%	2%
Third-party	1%	2%
Startups	0%	1%
Competitors	0%	0%

Internal Innovation Remains Critical

In a survey across 320 companies and 640 projects, respondents were asked which internal and external innovation sources were most important to their overall business and which produced their most successful (recent) innovation project. On balance, they deemed the internal sources—particularly central R&D, innovation labs, and dedicated business unit staff—more critical than external sources, such as co-innovation with customers and partnerships with startups.

brought into the company. There's no silver bullet, though; the circumstances will determine each company's approach.

It's essential to think through different models and develop several paths for bringing needed skills into the company. In some cases, a company will be able to source technological capabilities externally; in others, it may make sense to acquire (that is, "acquihire") startups. A third option may be to develop a build-run-transfer partnership. That arrangement will allow a technology business with the capabilities the company needs to build a dedicated team and manage it initially. Over time, the partner will transfer the team and all its work to the parent.

A Case in Point: Monsanto's Digital Portfolio

Some wise companies are learning to take a portfolio approach to innovation. Consider, by way of illustration, Monsanto, the American agrochemicals and agriculture company that became part of Germany's Bayer Group in 2018. Several years ago, Monsanto decided that data science would be critical to gain sustainable competitive advantage in its businesses. By gathering and studying farm data, it could not only manage its inventory more effectively but also help farmers decide which crops to plant, which seeds to use, how much fertilizer and water were required, and so on.

Monsanto therefore started the process of transforming itself from an agricultural biotechnology company into a data-science-driven organization. Digital agriculture, as the strategy is known, entails the mass collection of farm data through sensors attached to everything from tractors to water sources. All the data is fed through a digital platform set up by a service provider, whose algorithms display conditions on the farm and make specific recommendations.

Then-CEO Hugh Grant and the company's other senior leaders felt that given its size and ambitions, Monsanto should invest in the technologies to gather and study all the data underlying decision-making on farms. Because of the complexity of the work and the scarcity of digital skills inside the company, Monsanto first turned to external sources for key capabilities. It tied up with Amazon Web Services and Google Cloud for the big data and analytics infrastructure, Atomwise for AI technologies, AT&T for collecting large amounts of data, and governments and various experts for cybersecurity support.

Monsanto also built capabilities by working with, and acquiring, startups. Each year, it meets with around 250 startups, does 30 proofs of concept, and absorbs maybe five key technologies. In 2012, for instance, Monsanto purchased Precision Planting, which produces computer hardware and software that enables farmers to increase yields through more precise planting, for $210 million. The next year, Monsanto made an even larger acquisition, buying San Francisco–based Climate Corp., which provides weather forecasts for farmers based on modeling historic data, for $930 million.

To successfully tap more external innovation sources, Monsanto changed its innovation architecture. It built a data science center of excellence with a centralized data platform that is API- and microservices-driven. Monsanto adapted a software platform from Boston-based AI software company DataRobot, which allows agricultural experts to develop AI models without writing any code. Hundreds of models (a third of which are machine-learning-based) run on the platform to develop innovations for the company's supply chain, its commercial processes, and, of course, farmers.

Monsanto's new organizational constructs facilitate closer links with partners and greater absorption of key capabilities. The company delegates key employees to all the innovation projects it conducts with external partners in order to ensure active learning and knowledge transfer. From the get-go, Monsanto also internalized critical capabilities through training. Many employees re-skilled by taking Coursera and Data Camp courses. Over the past three years, Monsanto's data science community has grown from fewer than 200 people to more than 500, with many biologists and process chemists turning into data scientists.

Sums up Jim Swanson, Monsanto's CIO: "You have to look at your assets and where you need a partner. . . . For us, data and our scientific understanding of the data are tremendous assets. We've realized that internal capabilities are critical to our future, and we're investing in them. We're investing in the foundations of the networks and the modernization of our infrastructure. We've made tough decisions about the talent we need in digital, which is different than four years ago. We declared what was important for our future, and we've invested in the talent, the training, and we've been rigorous on how we measure, monitor, and advance. The more we do that, the more it accelerates our innovation. At the same time, we're also open about getting rid of the 'not invented here' mindset."

15

To Disrupt or Not to Disrupt?

Joshua Gans

The term *disrupt* has become synonymous with being an ambitious startup of any type. There's an almost cult-like devotion to the idea that becoming a disrupter is the best path to success—witness, for example, the annual TechCrunch Disrupt conference. But most studies of disruption have focused on the *disrupted*—why businesses that are seemingly at the top of their game suddenly find themselves in distress. In short, industry leaders are vulnerable to disruption when they are stuck in their profitable business model, finding themselves unable to see or respond to the mismatch between what they are offering and what current or future customers *actually* want. In almost every instance, disruption is precipitated by a new technological opportunity.

But even if market leaders in an industry are hamstrung in exploiting those new opportunities, can we take for granted that others—notably, new entrepreneurial entrants—will be able to do so? And even if they are able to seize those opportunities, should they? Disruption is a choice. But it's only one of many viable options for startups. Rather than single-mindedly heading down the path of would-be disrupter, new entrepreneurial companies can and should evaluate the trade-offs between disruption and other strategies. Doing so allows them to choose a strategy that is right for that startup, in that market, at that time, and to learn as the company commercializes its idea. To disrupt, or not to disrupt? That is a very important question. Here's how to think it through.

The Analysis

- The thinking in this article draws on the author's research studies with colleagues at MIT Sloan (Erin Scott, Scott Stern, Jane Wu, and Matt Marx) and Wharton (David Hsu).
- It also reflects work done by the author in *The Disruption Dilemma* (MIT Press, 2016).

A Tale of Two Startups

Even the early days of disruption saw glaring examples of alternative paths that could be chosen. In the late 1990s and early 2000s, the internet had just gone commercial, and there were numerous attempts to exploit it as a technological opportunity. One domain that garnered initial attention was the prospect for online grocery shopping. One company, in particular, stood out as a potential disrupter of grocery retailing. As we will see, things didn't really work out. What's more, as will be explained later, those lessons have not (yet) been learned.

Webvan was perhaps the quintessential dotcom company, enjoying a massive IPO of over $4 billion before going bankrupt just three years later, in 2001. During its years of operation, Webvan offered a unique and, in many ways, beloved service. Customers could log on to its website, do their entire grocery shopping online, and have it delivered to their door—for *less* than they would pay at the supermarket. Its advertisements highlighted the consumer pain point of waiting in store lines. The company's plan was to generate enough scale to use local distribution centers to ship goods to people and bypass supermarkets altogether, saving on stocking, rent, and, of course, bricks and mortar.

However, the plan didn't work: Webvan could not deliver goods at a cost that allowed the company to keep charging low prices. As it turned out, creating a new value chain for distributing goods to customers was expensive, involving massive investments in logistics and distribution centers. Unless customers ended up buying more groceries than before, Webvan would never reach a scale that justified those costs. Despite its value proposition, it wasn't able to attract enough consumers—even in the internet-savvy Bay Area—to generate economies of scale. And there were other issues, such as supplying products that needed refrigeration. All this meant that the would-be disrupter flamed out before supermarkets noticed a difference in their bottom line.

To use the internet for grocery shopping seemed like a good idea. But was the strategy of being a disrupter the right way to go? We know now that it wasn't (at least back then). Another entrepreneurial venture, Peapod, saw a similar opportunity but chose a different path. Founded by the Parkinson brothers in 1989, prior to the commercial internet, Peapod initially used computer networking to allow consumers to purchase groceries from supermarket chains such as Jewel, Kroger, and Safeway. In 1996, with the internet more freely available, Peapod set up a website and expanded its supermarket chain partnerships. The strategy was to hire people to shop at grocery outlets on behalf of Peapod's customers. Advertisements featured busy professionals—most often women—who didn't have time to do the grocery shopping. And Peapod charged a premium for the service.

Compared with Webvan, Peapod's strategy was decidedly nondisruptive. Supermarkets were its partners, not competitors to eventually be consigned to the dustbin of history. To be sure, Peapod attracted less funding and no financial exuberance, but it didn't need as much—its mission was not to construct an entirely new value chain but to slot itself into an existing one. And its customers were not those looking for a bargain, but those willing to pay a premium for convenience. In other words, Peapod positioned itself at the high end of the market rather than at a low end. Nothing it did was anywhere in the disrupter playbook.

Things worked out well for Peapod. It went from a successful IPO to growth, to flirting with some distribution assets before being acquired by Ahold—the owner of Stop & Shop—in 2000. It exists as a subsidiary of that company today.

A Tale of Two Other Startups

The path of a disrupter is not always a lucrative one. As part of its makeup, the disrupter chooses to take on established businesses, and sometimes an entire system, head-to-head. Competition is never easy and requires an aggressive, up-front investment. Partnering within the system appears to be an easier path, requiring fewer resources and incremental value. But it is far from clear that partnering is a path to sustained success.

That was surely apparent to Webvan's founder, Louis Borders, who also founded the eponymous Borders chain of bookstores. By the late 1990s, traditional bookstore chains started to see their sales challenged by a new entrant, Amazon.com. Amazon was founded in 1994 by Jeff Bezos, who moved to Seattle from Wall Street, not to sell books, but to take advantage of the opportunity presented by the internet. He chose books because he believed that they would be easy to ship without being damaged, consumers knew what they were paying for, and it was expensive for traditional brick-and-mortar retailers to stock a large variety of books. In Bezos's equation, variety was key; hence, the name Amazon to connote immense size.

Amazon was a disrupter by choice. It had its own website and sourced books independently of book retailers. When it entered, however, there had been a few precursors. For instance, in 1992 Charles Stack created Book Stacks Unlimited, a Cleveland-based outfit for dial-up book ordering. It soon offered a website, Books.com, that offered a large selection of titles but sourced its books from existing retailers. In other words, if Amazon was Webvan, Books.com was more like Peapod. By contrast, however, its life was relatively short; Books.com was acquired by an online player, Cendant, and ended up in the hands of Barnes & Noble.

It would be tempting to try to explain these disparate cases by pointing out failures in execution by Webvan and Book Stacks or by suggesting that the grocery and book markets were different in terms of the "right time" to exploit their respective opportunities as a disrupter. But the stories, I believe, carry another lesson: There is nothing inevitable about disruption, because there is *no compelling reason* when an entrepreneurial opportunity emerges to be a disrupter rather than something else. If anything, the lesson is that to be a disrupter, a company has to tailor *all* of its strategic choices toward that goal, as Amazon did. Similarly, Netflix successfully disrupted Blockbuster (and other Main Street video chains) in part because it always understood that it was creating an alternative value chain to video stores. It was never tempted to engage in halfway solutions that included physical drop-off and pick-up points (something the "vendor machine" rental operations such as RedBox attempted).

The conclusion? Choosing to be a disrupter should not be a startup's first choice. It's a hard road—much harder, longer, and resource-intensive than many new entrants realize. That doesn't mean there's not a viable path to disruption, but disruption should be a considered choice, and there are alternatives. An entrepreneurial startup should weigh each scenario carefully before going all in. Here's how to think through the right strategy for any particular circumstance.

The Disrupter's Choices

A convenient way to work through the problem involves compartmentalizing choices into four categories—technology, customer, organization, and competition. Being a disrupter requires particular orientations toward each of these categories, as I have explored in work with colleagues at MIT.[1]

First, consider the choice of *technology*. Clayton M. Christensen long distinguished between disruptive technologies (which perform worse today on metrics most consumers care about) and sustaining technologies (which do not). Most companies pursue sustaining technologies (such as modest improvements in an iPhone upgrade) as a way of retaining existing customers and keeping a healthy profit margin. The reason to choose a technology that is "worse" initially is its potential to outperform older technologies in the relatively near future. Moreover, disruptive technologies tend to be what established companies either are

A convenient way to work through the problem involves compartmentalizing choices into four categories – technology, customer, organization, and competition.

not good at or do not want to adopt for fear of alienating their customer base. In other words, the very existence of disruptive technologies represents an opportunity for startups.

Which brings us to the choice of *customer* for a disruptive entrepreneur. Christensen noted that, if you want to sell a product that underperforms existing products in some dimension (say, a laptop with less computing power), you need to find either a way of selling at a discount so that a lack of performance can be compensated for or a set of customers who do not strongly value that performance more than some other feature (for example, longer battery life). This was a struggle for Webvan. The company entered the general grocery business hoping to meet all customers' needs instead of seeking a more targeted base from which to build its internet retailing business.

Those low-end customers will establish a beachhead for the business. However, the startup cannot stop there, as Eric Ries has emphasized.[2] It needs to move rapidly up the performance technology curve to take advantage of its disruptive potential and grow. This requires rapid market-facing experimentation and rapid product performance improvement that will not only retain its newly acquired low-end customers but also allow it to compete for mainstream customers. A recent example is Soylent.com, which produces "pure nutritional need" food products. The company has experimented with different flavors as well as product types in response to market feedback.

Choosing to aim for mainstream customers, in turn, requires choosing an *organization* tailored to the hustle, market responsiveness, and capability investment that will allow for such growth. Amazon did this spectacularly well, right from the outset—gathering customer information and developing capabilities to predict demand for millions of different products.

The final choice to consider is what you want to *compete* with. Is your company headed down a disruptive path by adopting technology on a trajectory distinct from market leaders? Are you targeting customers you believe they aren't serving well? And have you built sufficient organizational capabilities to take advantage of that opportunity? If so, it wouldn't be surprising if you chose head-to-head competition with incumbents versus cooperating with them. Webvan could have chosen to slot itself in the existing value chain for groceries but did not. Amazon could have dealt directly with existing booksellers but did not. In each case, the incumbents eventually became targets.

But if all of these conditions are not clear—and desirable—an entrepreneur can choose alternative paths. One option is to become a value chain entrepreneur, partnering with existing market leaders. A startup can do this by slotting itself into the value chain and either becoming a customer of those incumbents or a supplier to them. Peapod chose to be, in effect, Stop & Shop's customer. But it was not an arm's-length relationship. The companies formed agreements that allowed Peapod to more smoothly run its online service and give customers access to existing supermarket products and prices. Taiwan-based electronics manufacturer Foxconn Technology has built its business on being a supplier of components and assembled products for Apple, Samsung, and others. Foxconn is not consumer-facing; instead, it works closely with designers to ensure it can deliver high-quality products efficiently.

How to Choose

Any entrepreneur with insight on how to exploit a new technological opportunity has many choices with respect to which strategy to use to bring that insight to market. Entrepreneurs can exploit technological opportunities via different paths, depending on their decisions about the four crucial strategic choices. But that's not to say that such choices are always clear-cut. In fact, inevitably there will be considerable uncertainty regarding which strategy is best and, indeed, whether there is one strategy that will turn out to be the best.

Given this, how can an entrepreneur decide whether disruption is the most appropriate path?

To expose the set of assumptions that might drive the success of a posited disruptive strategy, I would suggest that entrepreneurs put themselves through an adversarial process. First, they should outline the technology, customer, and organizational choices they would need to make in order to build a new business that could take on existing market leaders. In so doing, they need to ask themselves: Under what conditions will this path create value for identified customers? And under what conditions might an incumbent's competitive response be muted or delayed? An incumbent who is slow to respond is ideal for a would-be disrupter; an incumbent with a deep resource base and a quick reaction to a new threat can obliterate a would-be disrupter swiftly.

Having outlined a disruptive business plan, you should set that aside and draw up an alternative value chain plan. Is there a *different* path to success? Ask yourself: How can your company cooperate with existing market leaders to bring a technological opportunity to market? How would your company go about adding value to customers in the existing value chain or system? How can your company add to existing technologies—perhaps in a modular way? And can you build connections in the organization that would make it the preferred partner to incumbent businesses? You will need a clear statement of how your product adds value in existing value chains and under what conditions you will have sufficient bargaining power to capture some of that value.

The end result of this exercise is not one but two business plans—one of a disrupter and one not. You will then be in a position to choose.

What will guide that choice? In some situations, the choice may be easy. For instance, the entrepreneur may not be able to access the resources to undertake one of the plans. Or a plan may lack coherence—there may be no path from a targeted set of beachhead customers to generating market feedback and exploiting a more dynamic technological opportunity. In these cases, a plan can be easily discarded.

In other situations, perhaps mainly when there are valuable technological opportunities either way, both plans will look good, and the choice will come down to other factors. A financier may be attracted to one more than the other. Or the entrepreneur may have preexisting relationships with partners in the value chain that make one path more natural. Or the entrepreneur may simply identify with one path more than the other. Think of entrepreneurs such as Richard Branson who wanted to shake up markets: Even if they could have chosen other paths, they did not. That said, most entrepreneurs would be better off laying out their choices *before* making them.

Prepare to Pivot

Is disruption a binary choice? Does an entrepreneur need to know up front whether to pursue that path? In reality, the choice will probably have a dynamic component: If a strategy is found to be lacking, the entrepreneur can potentially switch paths. A shift in strategy from disruption to value chain or the reverse, without changing the core idea of the venture, is called a *pivot*. The difference between a successful pivot and a failed strategy of sticking with the wrong plan for too long often comes down to how well that company prepared for the possibility of Plan B.

Pivots can occur in either direction. For instance, strategy and innovation researcher Matt Marx and management scholar David Hsu have documented the case of Genentech, a biotech startup that was founded with a view of cooperating with existing pharmaceutical companies and licensing drug prospects to them.[3] Genentech did this with its breakthrough technique for producing synthetic insulin, which it licensed to Eli Lilly and Co. However, it also had aspirations to make its own pharmaceutical products. This involved garnering key regulatory skills and marketing capabilities. By licensing products and learning from licensees, Genentech was able to pivot beyond intellectual property development within a decade of its founding.

In other cases, startups might pivot from disruption to a value chain strategy. In research with Marx and Hsu,[4] I examined startup entry in the voice recognition software industry over an almost 50-year period. We found that a number of startups first entered and competed head-to-head with market leaders—presumably on a disruption path—but pivoted to becoming the incumbents' partners.

Ironically, we found that startups that pivoted away from disruption almost invariably did so because the technologies they developed were in fact disruptive. Why would a successful disruption lead anyone to pivot away from a disruptive strategy and toward a value chain opportunity? A technology disruption may provide a clear strategic path for a startup, but an incumbent might see the threat and opt to cooperate rather than compete with the entrepreneur. That, in turn, might end up as a win-win. The startup wouldn't need to fund an all-out battle with the incumbent but could find common ground for cooperating while enticing new customers with the improving technology. This means that even if entrepreneurs start out on a disruptive path, disruption might be forestalled as incumbents see that cooperating is in their interests. By adjusting their response to take advantage of the new entry, the incumbents can funnel the disruptive impact away from their own business.

Pivots can be planned (an entrepreneur may compete initially to show incumbents the value of cooperation) or unplanned (if a disruptive strategy is not effective and warrants a change). Either way, the possibility of pivoting means that any initial choice can be seen not as a final decision but as a potential continued learning opportunity. Just make sure you've walked through your pivot scenario in enough detail to know when it's time to execute it.

The Future

Disruption in an industry is a complex phenomenon. Market leaders might be vulnerable, but it takes others to make disruption happen. It is not a foregone conclusion. There is a viable alternative—a value chain approach.

The future of disruption likely depends not only on technological opportunities and their characteristics but also on the tools for experimentation and understanding that allow entrepreneurs' choices of technology, customer, organization, and competition to coalesce into a coherent whole. Disruption is an option. But there is a choice.

Contributors

Scott D. Anthony is a senior partner at growth strategy consultancy Innosight and coauthor of *Dual Transformation: How to Reposition Today's Business While Creating the Future* (Harvard Business School Publishing, 2017).

Didier Bonnet is an executive vice president at Capgemini Invent and a professor of strategy at IMD Business School in Lausanne, Switzerland.

Greg Brown is senior director of worldwide CAD business development at the global software company PTC.

Clayton M. Christensen, who passed away January 23, 2020, was Harvard Business School's Kim B. Clark Professor of Business Administration, an acclaimed author and teacher, and the world's foremost authority on disruptive innovation.

Michael A. Cusumano is the Sloan Management Review Distinguished Professor of Management at MIT Sloan School of Management.

Karen Dillon is a former editor of *Harvard Business Review* and coauthor of three bestselling books with Clayton M. Christensen.

Jeff Dyer is the Horace Beesley Professor of Strategy at Brigham Young University.

Sebastian K. Fixson is associate dean of innovation and the Marla M. Capozzi MBA '96 Term Chair of Design Thinking, Innovation, and Entrepreneurship at Babson College in Wellesley, Massachusetts.

Nathan Furr is a strategy and innovation professor at INSEAD.

Joshua Gans is the Jeffrey S. Skoll Chair of Technical Innovation and Entrepreneurship at the University of Toronto's Rotman School of Management and serves as chief economist in the Creative Destruction Lab.

Annabelle Gawer is Chaired Professor in Digital Economy at Surrey Business School at the University of Surrey. This article is adapted from *The Business of Platforms: Strategy in the Age of Digital Competition, Innovation, and Power* (HarperCollins, 2019).

Brian Halligan is CEO of HubSpot, a customer experience software platform for growing businesses. He's a coauthor of *Inbound Marketing: Attract, Engage, and Delight Customers Online* (John Wiley & Sons, 2014). This article was adapted from Halligan's keynote speech at HubSpot's Inbound 2019 event.

Nicole Helmer is a decision scientist at SAP, working at the intersection of customer experience, emerging technologies, and new product development.

Mike Hendron is an associate professor of entrepreneurship at Brigham Young.

Michael B. Horn, coauthor (with Bob Moesta) of *Choosing College* (Jossey-Bass, 2019), is the chief strategy officer at the Entangled Group, an education venture studio, and cofounder of the Clayton Christensen Institute, a nonprofit think tank. He has worked closely with some of the companies mentioned in this article, including several that have been clients of Entangled.

Marco Iansiti is the David Sarnoff Professor of Business Administration at Harvard Business School and heads its Technology and Operations Management Unit and its Digital Initiative. He is coauthor of *Competing in the Age of AI: Strategy and Leadership When Algorithms and Networks Run the World* (Harvard Business Review Press, 2020).

Rahul Kapoor is an associate professor of management at the Wharton School at the University of Pennsylvania in Philadelphia.

Thomas Klueter is an associate professor of entrepreneurship at IESE Business School at the University of Navarra in Barcelona.

Karim R. Lakhani is the Charles E. Wilson Professor of Business Administration at Harvard Business School and founder and codirector of Harvard's Laboratory for Innovation Science. He is coauthor of *Competing in the Age of AI: Strategy and Leadership When Algorithms and Networks Run the World* (Harvard Business Review Press, 2020).

Tucker J. Marion is an associate professor of technological entrepreneurship at Northeastern University's D'Amore-McKim School of Business in Boston.

Rita Gunther McGrath, a professor at Columbia Business School, is a globally recognized expert on strategy in uncertain and volatile environments. She is the author of *The End of Competitive Advantage* (Harvard Business Review Press, 2013) and *Seeing Around Corners* (Houghton Mifflin Harcourt, 2019).

Michael Putz is a strategy and business development executive with two decades of experience driving growth through disruptive innovation and business transformation.

Neil C. Thompson is a research scientist at the MIT Computer Science and Artificial Intelligence Lab (CSAIL) and the MIT Initiative on the Digital Economy.

Amy Webb is the founder of the Future Today Institute and professor of strategic foresight at the New York University Stern School of Business.

Max Wessel is chief innovation officer at SAP, responsible for technology research and product incubation efforts.

Ardine Williams is vice president of workforce development at Amazon's HQ2.

Yun Ye is a senior manager at Capgemini Invent in Shanghai.

David B. Yoffie is the Max and Doris Starr Professor of International Business Administration at Harvard Business School.

Notes

Chapter 2: How Leaders Delude Themselves about Disruption

1. O. Abbosh, M. Moore, B. Moussavi, et al., "Disruption Need Not Be an Enigma," Accenture, February 26, 2018, https://www.accenture.com/_acnmedia/pdf-72/accenture-disruptability-index-pov-final.pdf.

2. S. D. Anthony, A. Trotter, R. D. Bell, et al., "The Transformation 20: The Top Global Companies Leading Strategic Transformations," Innosight, September 2019, https://www.innosight.com/insight/the-transformation-20/.

3. "Interview with BlackBerry Co-CEO Jim Balsillie," CBC's *The Hour*, April 1, 2008, https://youtu.be/wQRcEObmSRM.

4. S. D. Anthony, C. G. Gilbert, M. W. Johnson, et al., "The Courage to Choose," in *Dual Transformation: How to Reposition Today's Business While Creating the Future* (Boston: Harvard Business Review Press, 2017).

5. M. D. Lord, S. W. Mandel, and J. D. Wager, "Spinning Out a Star," *Harvard Business Review* 80, no. 6 (June 2002): 115–121; R. Casadesus-Masanell and K. Elterman, "Walmart's Omnichannel Strategy: Revolution or Miscalculation?" Harvard Business School case no. 720-370 (Boston: Harvard Business School Publishing, 2019); J. Russell, "Walmart Sells Yihaodian, Its Chinese E-Commerce Marketplace, to Alibaba Rival JD.com," *TechCrunch*, June 21, 2016, https://techcrunch.com/2016/06/20/walmart-sells-yihaodian-its-chinese-e-commerce-marketplace-to-alibaba-rival-jd-com/; and D. B. Yoffie and D. Fisher, "Walmart Update, 2019," Harvard Business School case no. 719-504 (Boston: Harvard Business School Publishing, 2019).

6. D. Barton, J. Manyika, T. Koller, et al., "Where Companies with a Long-Term View Outperform Their Peers," McKinsey Global Institute, February 2017, https://www.mckinsey.com/featured-insights/long-term-capitalism/where-companies-with-a-long-term-view-outperform-their-peers.

7. M. J. Mauboussin and A. Rappaport, "Transparent Corporate Objectives—A Win-Win for Investors and the Companies They Invest In," *Journal of Applied Corporate Finance* 27, no. 2 (spring 2015): 28–33.

8. S. D. Anthony, C. G. Gilbert, M. W. Johnson, et al., "The Conviction to Persevere," *Dual Transformation: How to Reposition Today's Business While Creating the Future* (Boston: Harvard Business Review Press, 2017).

9. Harvard Business Review Staff, "The Best-Performing CEOs in the World 2018," *Harvard Business Review* 96, no. 6 (November–December 2018): 37–49.

10. "Leading Transformation: CEO Summit 2018," Innosight, August 2, 2018, https://www.innosight.com/insight/leading-transformation-lexington-ceo-summit-2018/.

11. S. D. Anthony, P. Cobban, R. Nair, et al., "Breaking Down the Barriers to Innovation," *Harvard Business Review* 97, no. 6 (November–December 2019): 92–101.

12. R. Kegan and L. L. Lahey, *An Everyone Culture: Becoming a Deliberately Developmental Organization* (Boston: Harvard Business School Publishing, 2016), 63.

13. M. Putz and M. E. Raynor, "Integral Leadership: Overcoming the Paradox of Growth," *Strategy & Leadership* 33, no. 1 (2005): 46–48.

14. J. Carter and R. Hougaard, "The Misconceptions of Mindfulness at Work," Greater Good Science Center, February 12, 2016, https://youtu.be/Iq6nSdNVuvE.

15. S. Abrams, "Mindfulness Is Aetna CEO's Prescription for Success," *HuffPost*, January 11, 2018, https://www.huffpost.com/entry/mindfulness-is-aetna-ceos-prescription-for-success_b_5a4bf577e4b0d86c803c7a1f.

16. T. Hindle, *Guide to Management Ideas and Gurus* (London: Profile Books, July 2008).

17. P. Wack, "Scenarios: Shooting the Rapids," *Harvard Business Review* 63, no. 6 (November 1985): 139–150.

18. C. Machmeier, "Mastering Digital Transformation with Mindfulness," SAP, September 3, 2018, https://news.sap.com/2018/09/mindfulness-training-master-digital-transformation/.
19. J. A. Quelch and C. Knoop, "Johnson & Johnson: The Promotion of Wellness," Harvard Business School case no. 514-112 (Boston: Harvard Business School Publishing, 2014).

Chapter 3: Overcoming the Innovator's Paradox

1. The term "innovator's paradox" is inspired by Nobel Prize–winning economist Kenneth Arrow's description of the information paradox, in which the seller of information can convince the buyer of its value only after it has been fully disclosed, at which point the information has already been transferred to the buyer, who hasn't paid for it. In an analogous manner, we argue that innovators struggle to convince the buyer (investors, customers, supporters) to support a new idea before the uncertainty associated with the idea has been resolved, at which point it is no longer a risk and thus support would be easy to come by.
2. For additional information on innovation capital, see J. Dyer, N. Furr, and C. Lefrandt, *Innovation Capital: How to Compete—and Win—Like the World's Most Innovative Leaders* (Boston: Harvard Business Review Press, 2019).
3. M. L. Martens, J. E. Jennings, and P. D. Jennings, "Do the Stories They Tell Get Them the Money They Need? The Role of Entrepreneurial Narratives in Resource Acquisition," *Academy of Management Journal* 50, no. 5 (October 2007): 1107–1132; and J. P. Cornelissen, J. S. Clarke, and A. Cienki, "Sensegiving in Entrepreneurial Contexts: The Use of Metaphors in Speech and Gesture to Gain and Sustain Support for Novel Business Ventures," *International Small Business Journal* 30, no. 3 (April 2012): 213–241.
4. C. Davenport, *The Space Barons: Elon Musk, Jeff Bezos, and the Quest to Colonize the Cosmos* (New York: PublicAffairs, 2018).
5. M. A. McDaniel and G. O. Einstein, "Bizarre Imagery as an Effective Memory Aid: The Importance of Distinctiveness," *Journal of Experimental Psychology: Learning, Memory, and Cognition* 12, no. 1 (January 1986): 54–65.

6. J. Bigelow and A. Poremba, "Achilles' Ear? Inferior Human Short-Term and Recognition Memory in the Auditory Modality," *PloS One* 9, no. 2 (February 26, 2014).
7. V. P. Seidel and S. O'Mahony, "Managing the Repertoire: Stories, Metaphors, Prototypes, and Concept Coherence in Product Innovation," *Organization Science* 25, no. 3 (May–June 2014): 691–712.
8. K. Weill, "Elon Musk Hyperloop Dreams Slam into Cold Hard Reality," *Daily Beast*, March 29, 2019, https://www.thedailybeast.com/elon-musk-hyperloop-dreams-slam-into-cold-hard-reality.
9. G. J. Stephens, L. J. Silbert, and U. Hasson, "Speaker-Listener Neural Coupling Underlies Successful Communication," *Proceedings of the National Academy of Sciences* 107, no. 32 (August 10, 2010): 14425–14430; U. Hasson, Y. Nir, I. Levy, et al., "Intersubject Synchronization of Cortical Activity During Natural Vision," *Science* 303, no. 5664 (March 12, 2004): 1634–1640; and T. Sharot, *The Influential Mind: What the Brain Reveals About Our Power to Change Others* (New York: Henry Holt, 2017).
10. P. J. Zak, "How Stories Change the Brain," *Greater Good*, December 17, 2018, https://greatergood.berkeley.edu/article/item/how_stories_change_brain.
11. D. Clark, *Alibaba: The House That Jack Ma Built* (New York: HarperCollins, 2016).
12. R. Garud, H. A. Schildt, and T. K. Lant, "Entrepreneurial Storytelling, Future Expectations, and the Paradox of Legitimacy," *Organization Science* 25, no. 5 (September–October 2014): 1479–1492.
13. L. A. Plummer, T. H. Allison, and B. L. Connelly, "Better Together? Signaling Interactions in New Venture Pursuit of Initial External Capital," *Academy of Management Journal* 59, no. 5 (October 2016): 1585–1604.
14. B. L. Hallen, "The Causes and Consequences of the Initial Network Positions of New Organizations: From Whom Do Entrepreneurs Receive Investments?" *Administrative Science Quarterly* 53, no. 4 (December 2008): 685–718.
15. R. B. Cialdini, "Scarcity: The Rule of the Few," in *Influence: The Psychology of Persuasion* (New York: HarperCollins, 2006).

16. B. L. Hallen and K. M. Eisenhardt, "Catalyzing Strategies and Efficient Tie Formation: How Entrepreneurial Firms Obtain Investment Ties," *Academy of Management Journal* 55, no. 1 (February 2012): 35–70.
17. Hallen and Eisenhardt, "Catalyzing Strategies and Efficient Tie Formation."
18. D. Kahneman and A. Tversky, "Choices, Values, and Frames," *American Psychologist* 39, no. 4 (April 1984): 341–350.

Chapter 4: A Crisis of Ethics in Technology Innovation

1. J. Stempel, "Judge Lets Facebook Privacy Class Action Proceed, Calls Company's Views 'So Wrong,'" Reuters, September 9, 2019, https://www.reuters.com/article/us-facebook-lawsuit-privacy/judge-lets-facebook-privacy-class-action-proceed-calls-companys-views-so-wrong-idUSKCN1VU2G2.
2. Building on a theory popularized by Kim B. Clark: C. M. Christensen, M. E. Raynor, and M. Verlinden, "Skate to Where the Money Will Be," *Harvard Business Review* 79, no. 10 (November 2001): 72–83.
3. C. Domonoske, "Federal Judge Extends Order Blocking 3D Gun Blueprints from Internet," NPR, August 27, 2018, https://www.npr.org/2018/08/27/642306105/federal-judge-extends-order-blocking-3-d-gun-blueprints-from-internet.
4. J. Wheatland, "I Was a Top Executive at Cambridge Analytica. It Taught Me a Tough Lesson about Public Trust," *Perspectives*, CNN Business, August 19, 2019, https://www.cnn.com/2019/08/19/perspectives/cambridge-analytica-lessons-julian-wheatland/index.html.

Chapter 5: Four Skills Tomorrow's Innovation Workforce Will Need

1. T. J. Marion and S. K. Fixson, *The Innovation Navigator: Transforming Your Organization in the Era of Digital Design and Collaborative Culture* (Toronto: University of Toronto Press, 2018).
2. C. Y. Baldwin, "Where Do Transactions Come From? Modularity, Transactions, and the Boundaries of Firms," Industrial and Corporate Change 17, no. 1 (February 2008): 155–195.

3. S. Rezvani and K. Monahan, "The Millennial Mindset: Work Styles and Aspirations of Millennials," Deloitte Greenhouse, 2017, https://www2.deloitte.com/content/dam/Deloitte/us/Documents/process-and-operations/us-cons-millennial-mindset.pdf; and "Millennials at Work: Reshaping the Workplace," PwC, 2011, https://www.pwc.de/de/prozessoptimierung/assets/millennials-at-work-2011.pdf.

4. D. Gelles and N. Kitroeff, "Boeing Pilot Complained of 'Egregious' Issue with 737 Max in 2016," *New York Times*, October 18, 2019, https://nyti.ms/2nZuPLp.

5. G. M. Gavetti, R. Henderson, and S. Giorgi, "Kodak and the Digital Revolution (A)," Harvard Business School case no. 705-448 (Boston: Harvard Business School Publishing, November 2004; rev. November 2005).

Chapter 6: Education, Disrupted

1. "Fast Facts," ASU GSV Summit, accessed November 18, 2019, http://www.asugsvsummit.com.

2. M. Weise, "Re-skilling Me Softly: Why Are American Companies So Bad at Re-skilling?" *Quartz*, October 1, 2019, https://qz.com/work/1715385/american-companies-have-a-hiring-problem/; see also "17th Annual Global CEO Survey," PwC, February 2019, https://www.pwc.com/gx/en/services/sustainability/ceo-views-sustainability-perspective.html.

3. C. Westfall, "New Survey: Nearly Half of Workers Unsatisfied with Learning and Development Programs," *Forbes*, October 8, 2019, https://www.forbes.com/sites/chriswestfall/2019/10/08/new-survey-workers-unsatisfied-with-learning-and-development-programs-training-leadership/.

4. P. Fain, "Amazon to Spend $700 Million on Training, Mostly Outside College," *Inside Higher Ed*, July 12, 2019, https://www.insidehighered.com/quicktakes/2019/07/12/amazon-spend-700-million-training-mostly-outside-college.

5. D. Lederman, "Online Education Ascends," *Inside Higher Ed*, November 7, 2018, https://www.insidehighered.com/digital-learning/article/2018/11/07/new-data-online-enrollments-grow-and-share-overall-enrollment.

6. A. D. Salisbury, "As Pressure to Upskill Grows, Five Models Emerge," *Forbes*, October 28, 2019, https://www.forbes.com/sites/allisondulinsalisbury/2019/10/28/as-pressure-to-upskill-grows-5-models-emerge/.

7. S. Caminiti, "AT&T's $1 Billion Gambit: Retraining Nearly Half Its Workforce for Jobs of the Future," CNBC, March 13, 2018, https://www.cnbc.com/2018/03/13/atts-1-billion-gambit-retraining-nearly-half-its-workforce.html.

8. M. E. Echols, *ROI on Human Capital Investment*, 2nd ed. (Littleton, MA: Tapestry Press, 2005).

9. M. Murphy, "Why New Hires Fail (Emotional Intelligence vs. Skills)," *Leadership IQ* (blog), June 22, 2015, https://www.leadershipiq.com/blogs/leadershipiq/35354241-why-new-hires-fail-emotional-intelligence-vs-skills.

10. "Nearly Three in Four Employers Affected by a Bad Hire, According to a Recent CareerBuilder Survey," CareerBuilder, December 7, 2017, http://press.careerbuilder.com/2017-12-07-Nearly-Three-in-Four-Employers-Affected-by-a-Bad-Hire-According-to-a-Recent-CareerBuilder-Survey.

11. S. A. Ambrose, M. W. Bridges, M. DiPietro, et al., *How Learning Works: Seven Research-Based Principles for Smart Teaching* (San Francisco: John Wiley & Sons, 2010).

Chapter 8: From Disruption to Collision

1. R. Molla, "American Consumers Spent More on Airbnb Than on Hilton Last Year," *Vox*, March 25, 2019, https://www.vox.com/2019/3/25/18276296/airbnb-hotels-hilton-marriott-us-spending.

2. For an outline of the challenges faced by traditional corporations, see A. D. Chandler, Jr., *Scale and Scope: The Dynamics of Industrial Capitalism* (Cambridge, MA: Belknap Press, 1990).

3. While the scaling behavior of algorithms will vary with algorithm type and implementation, a basic rule of thumb is that the precision of an algorithm will scale with the square root of the number of data points.

4. C. Farronato and A. Fradkin, "The Welfare Effects of Peer Entry in the Accommodation Market: The Case of Airbnb," working paper 24361, National Bureau of Economic Research, Cambridge, Massachusetts, February 2018.

5. For an instruction manual on traditional retail businesses that can be dismantled through a digital operating model, see M. Zeng, *Smart Business: What Alibaba's Success Reveals About the Future of Strategy* (Boston: Harvard Business Review Press, 2018).

6. RealNetworks, originally known as Progressive Networks, was founded in 1994 by Rob Glaser.
7. M. E. Conway, "How Do Committees Invent?" *Datamation* 14, no. 5 (April 1968): 28–31.
8. R. Henderson and K. B. Clark, "Architectural Innovation: The Reconfiguration of Existing Product Technologies and the Failure of Established Firms," *Administrative Science Quarterly* 35, no. 1 (March 1990): 9–30.
9. This is not surprising, given that both Henderson and Clark were on Christensen's thesis committee at Harvard Business School.
10. C. M. Christensen and R. S. Rosenbloom, "Explaining the Attacker's Advantage: Technological Paradigms, Organizational Dynamics, and the Value Network," *Research Policy* 24, no. 2 (March 1995): 233–257.
11. According to Christensen, Raynor, and McDonald, "'Disruption' describes a process whereby a smaller company with fewer resources is able to successfully challenge established incumbent businesses. Specifically, as incumbents focus on improving their products and services for their most demanding (and usually most profitable) customers, they exceed the needs of some segments and ignore the needs of others." See C. M. Christensen, M. E. Raynor, and R. McDonald, "What Is Disruptive Innovation?" *Harvard Business Review* 93, no. 12 (December 2015): 44–53.

Chapter 9: The Future of Platforms

1. Based on public stock market valuations on January 22, 2020: Apple, $1.4 trillion; Microsoft, $1.3 trillion; Alphabet/Google $1 trillion; Amazon, $937 billion; Facebook, $633 billion; Alibaba, $602 billion; Tencent, $483 billion.
2. "The Crunchbase Unicorn Leaderboard," *TechCrunch*, accessed July 2017, https://techcrunch.com/unicorn-leaderboard/ (no longer available).
3. H. Somerville and P. Lienert, "Inside SoftBank's Push to Rule the Road," Reuters, April 12, 2019, https://www.reuters.com/article/us-softbank-autos-investments-insight/inside-softbanks-push-to-rule-the-road-idUSKCN1RO049.

4. R. Verger, "Someday, You Might Subscribe to a Self-Driving Taxi Service, Netflix-Style," *Popular Science*, March 15, 2018, https://www.popsci.com/lyft-subscription-self-driving-car/.
5. C. Mims, "How Self-Driving Cars Could End Uber," *Wall Street Journal*, May 7, 2017, https://www.wsj.com/articles/how-self-driving-cars-could-end-uber-1494154805.
6. J. Palmer, "Here, There, and Everywhere: Quantum Technology Is Beginning to Come into Its Own," *Economist*, March 9, 2017, https://www.economist.com/technology-quarterly/2017/03/09/quantum-technology-is-beginning-to-come-into-its-own.
7. "List of Companies Involved in Quantum Computing or Communication," Wikipedia, accessed May 26, 2018, https://en.wikipedia.org/wiki/List_of_companies_involved_in_quantum_computing_or_communication.
8. S. Aaronson, "Why Google's Quantum Supremacy Milestone Matters," *New York Times*, October 30, 2019, https://nyti.ms/2BVxMjj.

Chapter 10: The Experience Disrupters

1. J. J. Roberts, "IBM Tops 2018 Patent List as AI and Quantum Computing Gain Prominence," *Fortune*, January 7, 2019, https://fortune.com/2019/01/07/ibm-tops-2018-patent-list-as-ai-and-quantum-computing-gain-prominence/.
2. D. Muller, "Carvana Debuts as No. 8 on Used Ranking," *Automotive News*, April 22, 2019, https://www.autonews.com/used-cars/carvana-debuts-no-8-used-ranking.
3. L. Smiley, "Stitch Fix's Radical Data-Driven Way to Sell Clothes—$1.2 Billion Last Year—Is Reinventing Retail," *Fast Company*, February 19, 2019, https://www.fastcompany.com/90298900/stitch-fix-most-innovative-companies-2019.

Chapter 11: The New Disrupters

1. C. M. Christensen, *The Innovator's Dilemma: When New Technologies Cause Great Firms to Fail* (Boston: Harvard Business School Press, 1997).
2. M. E. Porter, *Competitive Strategy: Techniques for Analyzing Industries and Competitors* (New York: Free Press, 1980).

3. C. M. Christensen, M. E. Raynor, and R. McDonald, "What Is Disruptive Innovation?" *Harvard Business Review* 93, no. 12 (December 2015): 44–53.
4. W. A. Sahlman and H. H. Stevenson, "Capital Market Myopia," *Journal of Business Venturing* 1, no. 1 (winter 1985): 7–30.
5. C. M. Christensen, T. Hall, K. Dillon, et al., *Competing Against Luck: The Story of Innovation and Customer Choice* (New York: HarperBusiness, 2016).
6. C. M. Christensen and D. C. M. van Bever, "The Capitalist's Dilemma," *Harvard Business Review* 92, no. 6 (June 2014): 60–68.
7. W. Lazonick, "The Curse of Stock Buybacks," *American Prospect*, summer 2018, 34–38.
8. Christensen et al., "What Is Disruptive Innovation?"
9. T. McEnery, "Like a Body on Life Support Fluttering Its Eyelids, General Electric Releases Quarterly Results," *DealBreaker*, July 31, 2019, https://dealbreaker.com/2019/07/general-electric-post-still-alive-results-for-the-quarter.

Chapter 13: The Uncertainty Factor

1. C. M. Christensen, M. E. Raynor, and R. McDonald, "What Is Disruptive Innovation?" *Harvard Business Review* 93, no. 12 (December 2015): 44–53.
2. R. Kapoor and T. Klueter, "Decoding the Adaptability-Rigidity Puzzle: Evidence From Pharmaceutical Incumbents' Pursuit of Gene Therapy and Monoclonal Antibodies," *Academy of Management Journal* 58, no. 4 (August 2015): 1180–1207; R. Kapoor and T. Klueter, "Progress and Setbacks: The Two Faces of Technology Emergence," Research Policy 49, no. 1 (February 2020): article no. 103874; R. Kapoor, T. Klueter, and J. M. Wilson, "Challenges in the Gene Therapy Commercial Ecosystem," *Nature Biotechnology* 35, no. 9 (September 2017): 813–815; and R. Kapoor and T. Klueter, "Organizing for New Technologies," *MIT Sloan Management Review* 58, no. 2 (winter 2017): 85–86.
3. T. A. Brennan and J. M. Wilson, "The Special Case of Gene Therapy Pricing," *Nature Biotechnology* 32, no. 9 (September 2014): 874–876.
4. E. Warner, "Goodbye Glybera! The World's First Gene Therapy Will Be Withdrawn," *Labiotech.eu*, April 20, 2017, https://www.labiotech.eu/medical/uniqure-glybera-marketing-withdrawn/.

5. R. Kapoor and N. R. Furr, "Complementarities and Competition: Unpacking the Drivers of Entrants' Technology Choices in the Solar Photovoltaic Industry," *Strategic Management Journal* 36, no. 3 (February 2014): 416–436.
6. J. Eilperin, "Why the Clean Tech Boom Went Bust," *Wired*, January 20, 2018, https://www.wired.com/2012/01/ff_solyndra/.
7. L. Noel and B. K. Sovacool, "Why Did Better Place Fail? Range Anxiety, Interpretive Flexibility, and Electric Vehicle Promotion in Denmark and Israel," *Energy Policy* 94, no. 7 (July 2016): 377–386.
8. J. Eklund and R. Kapoor, "Pursuing the New While Sustaining the Current: Incumbent Strategies and Firm Value During the Nascent Period of Industry Change," *Organization Science* 30, no. 2 (March–April 2019): 383–404.
9. D. Ferris, E. Klump, and D. Kahn, "Why NRG's Green Crusade Faltered," *Energywire*, March 7, 2016, https://www.eenews.net/energywire/2016/03/07/stories/1060033507; and S. Lacey, "NRG Fully Exits the Home Solar Installation Business," *Greentech Media*, February 15, 2017, https://www.greentechmedia.com/articles/read/nrg-sheds-the-final-remnants-of-its-home-solar-business.

Chapter 15: To Disrupt or Not to Disrupt?

1. J. Gans, E. L. Scott, and S. Stern, "Strategy for Start-Ups," *Harvard Business Review* 96 (May–June 2018): 44–51; and J. S. Gans, S. Stern, and J. Wu, "Foundations of Entrepreneurial Strategy," *Strategic Management Journal* 40, no. 5 (May 2019): 736–756.
2. E. Ries, "The Lean Startup: How Today's Entrepreneurs Use Continuous Innovation to Create Radically Successful Businesses" (New York: Currency, 2011).
3. M. Marx and D. H. Hsu, "Strategic Switchbacks: Dynamic Commercialization Strategies for Technology Entrepreneurs," *Research Policy* 44, no. 10 (December 2015): 1815–1826.
4. M. Marx, J. S. Gans, and D. H. Hsu, "Dynamic Commercialization Strategies for Disruptive Technologies: Evidence from the Speech Recognition Industry," *Management Science* 60, no. 12 (December 2014): 3103–3123.

Index